THE NEW SCIENCE OF CONSCIOUSNESS

THE NEW SCIENCE OF CONSCIOUSNESS

Exploring the Complexity of Brain, Mind, and Self

PAUL L. NUNEZ

Prometheus Books

59 John Glenn Drive
Amherst, New York 14228

Published 2016 by Prometheus Books

Cover design by Jeff Schaller
Cover design © Prometheus Books
Cover image © iStockphoto.com/airiesummer

Inquiries should be addressed to
Prometheus Books
59 John Glenn Drive
Amherst, New York 14228
VOICE: 716–691–0133
FAX: 716–691–0137
WWW.PROMETHEUSBOOKS.COM

20 19 18 17 16   5 4 3 2 1

Library of Congress Cataloging-in-Publication Data

Names: Nunez, Paul L.
Title: The new science of consciousness : exploring the complexity of brain, mind, and
    self / by Paul L. Nunez, Ph.D.
Description: Amherst, New York : Prometheus Books, 2016. | Includes bibliographical
    references and index.
Identifiers: LCCN 2016017939 (print) | LCCN 2016019620 (ebook) |
    ISBN 9781633882195 (hardcover) | ISBN 9781633882201 (ebook)
Subjects: LCSH: Consciousness—Popular works. | Brain—Popular works. |
    Neurosciences—Popular works.
Classification: LCC QP411 .N856 2016 (print) | LCC QP411 (ebook) | DDC 612.8—dc23
LC record available at https://lccn.loc.gov/2016017939

Printed in the United States of America

CONTENTS

PREFACE

This book explains, in layman's language, a new approach to the science of consciousness based on collaborations between neuroscientists and complexity scientists. The new science goes well beyond traditional cognitive science and simple networks, which provide the usual focus of artificial-intelligence research. The new partnerships involve many subfields of neuroscience, physics, psychology, psychiatry, philosophy, and more. This cross-disciplinary approach aims to reveal fresh insights into the major unsolved challenge of our age: the origins of awareness, especially our self-awareness.

What causes autism, schizophrenia, and Alzheimer's disease? How does our unconscious influence our actions? How does human consciousness differ from the apparent consciousness of lower animals? Do we enjoy genuine free will or are we slaves to unconscious systems? Are conscious computers theoretically possible? Above all, how can the interactions of a hundred billion nerve cells lead to the mysterious condition called consciousness? To tell this compelling story, I recruit a number of analogues and metaphors, showing how brain behavior can be compared to the collective behaviors of other large-scale systems. The world human social system provides just one example in which novel features like wars and economic depressions emerge, features that are absent from the small parts. Emergent systems of many kinds interact and form relationships with lower levels of organization and the surrounding environment, thereby suggesting compelling models for complex brain functions. These studies raise the *Big Question*—do minds emerge from brains? Or is something more involved?

Since my early career move from engineering and theoretical physics to neuroscience, my professional life has focused on the "easy problem" of consciousness—exploring the relationships between brain and mind activity, the so-called signatures of consciousness. These experimental measures of brain activity, like electroencephalography (EEG) and functional magnetic resonance imaging (fMRI), are critically important to brain science. By contrast, the "hard problem," the how and why of our conscious awareness, presents a much more profound puzzle. Scientists and philosophers disagree as to whether the hard problem can ever be solved; some even deny that such problem exists. Like many scientists and nonscientists alike, I have a long-running fascination with the mystery of consciousness, which motivates my interests in several related scientific fields.

My earlier ventures into consciousness study resulted in *Brain, Mind, and the Structure of Reality*, published in 2010. While mostly well received, some readers found the material a bit too technical for their tastes. This new book extends the ideas presented in 2010 but, at the same time, aims to serve a wider readership by substantially softening the technical content and mostly avoiding mathematics. Not a single equation occupies these pages, excepting a few in the endnotes. Readers of the earlier book will note some duplication, aimed to ensure that even those unfamiliar with brain science can safely begin here. My approach rests on the following conceptual framework that enjoys full consistency with mainstream science: (1) Brains and minds are correlated; that is, many consciousness signatures have been discovered. (2) Brains are genuine complex systems; more complex than most other living things. (3) Brains, like other complex systems, consist of nested hierarchies of subsystems that operate at different levels of organization (spatial scales). (4) Consistent with this picture, signatures of consciousness are observed over a wide range of scales. (5) Multiple conscious, unconscious, and semi-conscious entities coexist within each human brain. (6) Interactions between these subsystems contribute substantially to making the human brain "human."

This framework supports an idea labeled the *multiscale conjecture*, which posits that consciousness manifests at multiple levels of brain organization, and no single scale need be special. Thus, the various dynamic patterns of information observed as consciousness signatures, as measured with different scientific methods, may all contribute to the mind. In this view, consciousness is rooted in the dynamic patterns of multiple interacting scales. Although fully consistent with mainstream science, the multiscale conjecture allows room for both materialistic and nonmaterialistic interpretations. Hence, I argue that materialism and dualism are not as distinct as many think. Some aspects of dualism are shown to be fully consistent with modern science.

The consciousness challenge is approached with questions about a category beyond ordinary information, that is, *ultra-information*, defined broadly to include ordinary information, hidden physical processes, and consciousness. Thoughts, emotions, self-awareness, memory, and the contents of the unconscious are, by definition, categories of ultra-information whether or not these mental processes also involve ordinary information. This idea may sound rather abstract, but it is consistent with modern physics, which tells us that some kinds of information are fundamentally unknowable. For example, all interpretations of quantum mechanics rely on the existence of some sort of hidden reality or "shadow world" that we can never observe directly, but that nevertheless influences the familiar world of our senses. By definition, this hidden world contains ultra-information.

In order to serve a general readership, my storytelling adopts a style counter to typical scientific books or journal articles—one may say that I aim for two degrees of separation from such technical material. The endnotes section is relatively short, with emphasis on review articles and books. The scientists who have contributed to this book's intellectual framework number in the thousands, but only a tiny fraction is cited here. Furthermore, phrases like "Professor A found X, but on the other hand, Professor B found Y" are largely absent; rather, an averaged-over consensus is mostly employed for easy reading. Some may be offended by this style because scientists tend to follow a familiar dictum, "the devil is in the details." However, many details are often missing in popular accounts, as is the case here. I offer my apologies to the scientists and philosophers whose important contributions are oversimplified or perhaps not cited at all. I trust that these extra efforts to simplify the narrative contribute to the reader's comprehension of this mind-expanding topic.

What about the controversial idea that quantum mechanics underpins consciousness in some way? This putative connection has been discussed in many publications, including my 2010 book, and is not examined here in detail. However, I do outline the postulate labeled *RQTC*, the conjecture that relativity, quantum mechanics, and thermodynamics *may* somehow play an essential role in the theater of consciousness. I offer no conclusive evidence, only hints of possible connections. Nevertheless, if the study of consciousness boils down to a more-general study of reality, as I firmly believe, the worldviews provided by modern physics must be seriously considered along with evidence from brain science.

Special thanks to Ed Kelly, the major reviewer, who read several drafts of the entire book; his comments and criticisms were extensive and insightful. Thanks also to the following friends, family, and colleagues for their helpful input: Xavier Alvarez, Bernie Baars, György Buzsáki, David Chalmers, Todd Feinberg, Doug Fields, Scott Kelso, Lester Ingber, Cindy McKnight, Kirsty Nunez, Lisa Nunez, Michael Nunez, Michelle Nunez, Shari Nunez, Richard Silberstein, Ramesh Srinivasan, Lawrence Ward, Brett Wingeier, and Jim Wright.

Chapter 1

INTRODUCTION TO MIND AND BRAIN

1.1 THE DEEP MYSTERY OF CONSCIOUSNESS

The challenge of consciousness is uniquely curious. I am conscious, but what about you? We have no direct experience of any consciousness other than our own. We infer that others share similar kinds of internal experiences mainly because they look, behave, and talk as we do. These similarities encourage the belief that others are aware of their own existence, thoughts, and environment just as we are. But how much do we really know about the internal experiences of others? How do you experience a Mozart concerto, a bungee jump, a cockroach in your kitchen, or a nude person surprising you in an elevator? What about your dog or cat—what kinds of internal experiences do they have? Or, in the colorful words of one philosopher, "What is it like to be a bat?"

The word *self* in this book's subtitle suggests that even if consciousness emerges from brain complexity as believed by most scientists, the very existence of the phenomenon of self-awareness remains a deep mystery. The title also suggests that self-awareness involves developments beyond mind; some lower animals can have "minds" but apparently lack a sense of "self." In other words, lower animals probably have *primary consciousness*, an awareness of their current environment. They may, however, lack the *higher consciousness* of higher animals, that is, those with awareness of self and its relation to the past, present, and future. If this distinction is indeed valid, just where do we draw the line separating higher from lower animals? Most of us think that dolphins, chimps, and dogs are conscious. But to what extent are they conscious, and how are humans special?

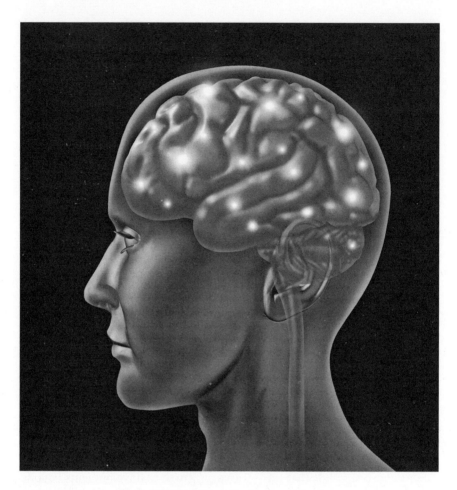

Fig. 1-1. © Can Stock Photo Inc. / bluering.

The phenomenon of consciousness is the major unsolved challenge of our age. Not only do we not have answers, we are often unsure of the right questions. This is not to say that the study of consciousness lies outside the province of science. The consciousness challenge may be divided into two parts, the *easy problem* and the *hard problem*. The "easy problem" is concerned with the various electrical or chemical measures of brain functions that occur in different brain states. These measures, called *consciousness correlates* or *signatures of consciousness*, are indicated symbolically in figure 1-1, where the lighted regions might indicate any one of several kinds of brain activity. For example, what happens to blood-oxygen levels in different brain regions when we recognize a face or carry out some task? Or, how do electrical patterns over the brain change between waking and sleeping or with various depths of anesthesia? How do brain injuries or diseases that afflict certain brain structures change our consciousness? What causes autism, schizophrenia, or Alzheimer's disease? And the really big question—how can the massive interactions between nerve cells (*neurons*), shown in figure 1-2, determine our thoughts and behaviors?

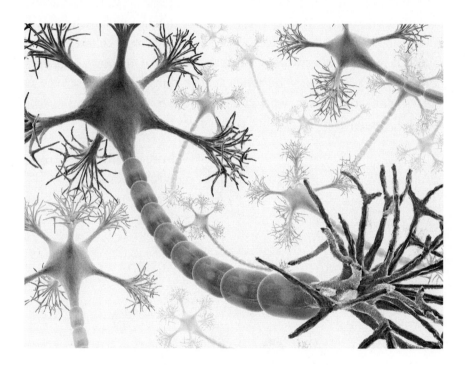

Fig. 1-2. © Can Stock Photo Inc. / iDesign.

No one questions that these studies of consciousness correlates represent legitimate science. By contrast, the hard problem is concerned with the very existence of the amazing phenomenon of conscious experience. Many scientists do not consider the hard problem to be a scientific problem at all; they would leave this issue at the feet of philosophers. We will look into this controversy from several perspectives in the following chapters. Suffice it to say, our scientific knowledge about how brains work grows every day, yet a central theme that permeates our story is the delicate balance between knowledge and ignorance: *how much we know, but how little we know of consciousness*.

Brains are "complex" in the manner understood in the field of *complexity science*, an exciting new approach adopted to study everything from social systems to ecology to economics to weather patterns. Plausible assumptions about the underlying causes for various healthy and disease states of brains are suggested by analogy to other complex systems that are better understood and more easily visualized, for example, human and animal social systems. Human social networks interact with each other in many complex ways; they are also embedded within larger cultures that act *top-down* on the local networks. The profound idea of *top-down influences across multiple levels of organization* applies to many areas of science as well as our everyday lives. We will return to this issue many times in later chapters, but the following outline summarizes the general idea.

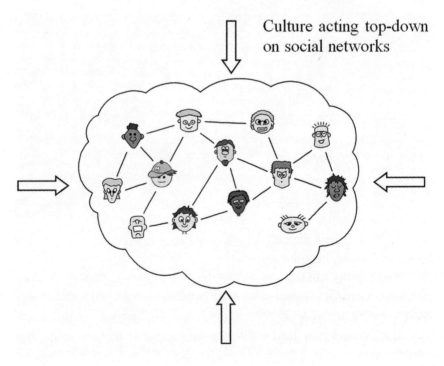

Culture acting top-down on social networks

Fig. 1-3.

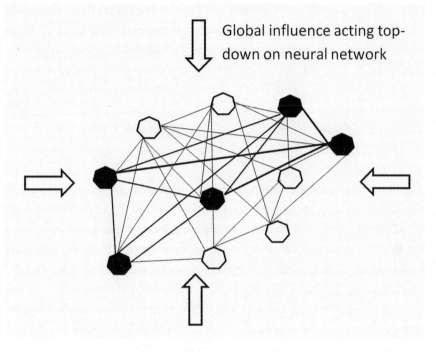

Fig. 1-4.

Our interactions with friends on religious, political, financial, and other issues are likely to be substantially influenced by larger systems like the mass media or the culture of our local geographic region, as indicated in figure 1-3. The arrows, both vertical and horizontal, indicate top-down influences across levels of organization. Are you a Democrat or a Republican? Are you a born-again Christian or an atheist? Regardless of our beliefs, the activities of any of our many overlapping social networks are likely to look quite different in Texas than in California. To take a more-obvious example, imagine that one of your social networks in the United States were to be suddenly transported to Saudi Arabia. The Saudi laws and customs would provide major top-down influences. Ask your wife to drive you to the airport? Forget it. Meet with friends to have drinks at a local restaurant? Don't even think about it. Such top-down social interactions provide useful and intriguing metaphors for the brain's neural networks, which are similarly influenced by top-down effects. As discussed in later chapters, top-down influences on neural networks originate from both inside and outside the brain, as indicated in figure 1-4. With such metaphors in mind, one might label this approach to brain science as *neuron sociology*.

1.2 A PERSONAL PERSPECTIVE ON SCIENCE, PHILOSOPHY, AND RELIGION

The scientific material, interpretations, and opinions in this book are biased by my educational, professional, and other life experiences. This is, of course, true of all books and their authors, but such influences can be masked by the author's desire to appear unbiased. This book addresses both the "easy" and "hard" problems of consciousness. Most agree that the so-called easy problem, finding brain correlates of conscious states, lies well within scientific purview. However, the hard problem is considered by many to lie outside of science, perhaps to be left to philosophy or religion. In contrast to most scientific publications, religious issues are faced in this book when they appear to overlap the consciousness challenge. After all, much of humanity follows some sort of faith, and conflicts between science and religion often surface when addressing the hard problem. Not surprisingly, religious persons typically favor explanations that support their faith. In response to such bias, scientists and their followers sometimes overreact by opposing ideas perhaps mainly because they "smell" a little too religious. Biases on one side can beget counter biases and so forth. With this background in mind, our goal here is to avoid all aspects of *epistemophobia*, a fancy label for the irrational fear of knowledge, regardless of its nature. We shall be entirely free to offend the faithful, the scientists, or anybody else when the evidence demands it.

Our aim here is a friendly discussion of the consciousness challenge with open minds. For this reason, it seems like a good idea for me to reveal my "tribal membership" at the outset. My fellow tribal members include scientists, religious persons, agnostics, atheists, and others with differing worldviews and from many walks of life. Our religious members differ from the dogmatically religious in making no claim of access to *personal truths* that automatically trump the beliefs of others. Our tribe rejects the idea that any mathematician, philosopher, scientist, rabbi, pope, priest, imam, pastor, Torah, Qur'ān (Koran), or Bible is immune from close scrutiny or criticism. We reject seduction by "guru wisdom." We believe that Truth is approached in a series of successive approximations, and at no point can final and complete accuracy be claimed. We employ the scientific method as our favorite tool, employed in order to achieve better approximations to Truth. But we don't automatically reject nor are we hostile to nonscientific beliefs like faith, intuition, speculation, and so forth, as long as they are properly acknowledged as such.

Consistent with my tribe's philosophy, I strongly support the scientific study of consciousness but reject extreme versions of scientism that place science at the pinnacle of all forms of human knowledge and experience, necessarily taking primacy over ethics, philosophy, religion, and humanistic views of reality. We hold that no field of knowledge should be insulated from scientific rigor when application of genuine scientific methods is possible. However, we must be ready to expose pseudoscientific babble and imposters masquerading as genuine science. Some prominent philosophers and scientists have implied that we are nothing but a collection of nerve cells (neurons), essentially robots lacking free will. Are our conscious "free choices" really nothing but illusions? In this view, consciousness is nothing but an accidental product of physical processes. But such views, while often promoted as the most "scientific," may actually be inconsistent with genuine twenty-first-century science.

1.3 WHAT'S SO "NEW" ABOUT THE SCIENCE OF CONSCIOUSNESS?

Modern brain science is very much concerned with the idea of *emergence*. This is a process whereby new large-scale features arise through the actions of smaller entities that do not possess these features in isolation. The important scientific term *scale* refers to various levels of organization in systems of nearly any kind. In both philosophy and science, *emergence* refers to novel holistic or "global" properties that arise in complex systems from relatively simple interactions within smaller-scale systems. Nations emerge from the collective interactions of individuals. The large protein molecules necessary for life have important properties that may be impossible to anticipate from the fundamental features of their atomic building blocks. In meteorology, tornadoes emerge from simpler interactions of small air masses. In some materials, superconductivity emerges from quantum interactions between atoms and electrons.

The general idea of the *top-down influence of emergent systems on smaller systems* is indicated in figure 1-5. In the philosophical or scientific position known as *strong reductionism*, the importance of emergent properties of higher-level systems acting top-down on lower-level systems is mostly discounted. Strong reductionism is nicely illustrated by the *great man theory*, in which history is explained by the impact of great men; that is, influential individuals who, due to charisma, wisdom, money, or political skills, employed their influence (bottom-up) in a way that provided a decisive large-scale historical impact. The counterargument holds that such great men are products of their societies, and that their actions would be impossible without the existing social and political conditions acting top-down. Wars, religions, and national economic and political policies are large-scale phenomena that act top-down on individuals at small scales, who then act bottom-up on the larger scales, as in the prominent examples of Jesus, Darwin, Marx, Einstein, Hitler, and Osama bin Laden. Modern complexity science explicitly recognizes such *circular causality*; that is, interactions across multiple levels of organization in both directions.

Such interactions can occur across many levels of organization, which are referred to as *spatial scales* in scientific circles. They can result in emergent systems, as demonstrated by figure 1-5, where the vertical arrows demonstrate circular causality, that is, the combined up-and-down interactions across spatial scales. The lowest-level system shown at the bottom consists of some kind of interacting entities. These can be almost anything; let's say they represent DNA, protein, and other molecules. Interactions between these molecules and with the external environment cause the (bottom-up) emergence of living cells. In this example, we may call the cellular level the "Level I emergent system." A single cell like the neuron may contain a hundred billion interacting molecules. It seems to act much like a natural supercomputer, an information-processing and replicating system of enormous complexity that can act down on the molecular scale and act up on the "Level II emergent system." Let a lot of cells interact in the right manner and in the right environment, and a human being will emerge. In this case, the emergent system is a person who can produce many top-down influences on the two lower levels, like when he eats a pizza, runs a race, visits his local bar, or hops in bed with his partner. A more-accurate picture of living systems would include intermediate emergent systems, for example, organ systems between the levels of cells and persons.

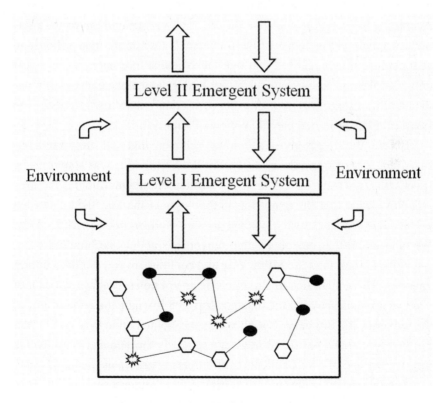

Fig. 1-5.

Many other examples come to mind. We might, for example, consider the lowest-level entities to be individual persons, with Levels I and II representing corporations and national populations, respectively. Lower-level social systems or individuals may act bottom-up to facilitate major political shifts, economic depressions, wars, and so forth. In the United States, the super wealthy can generate major bottom-up influences by buying and controlling newspapers and TV stations, donating to political campaigns, funding lobbyists, and so forth. The resulting large-scale systems like political action committees, corporate lobbies, and governments then act top-down at multiple lower levels, perhaps by writing or passing laws that encourage even more wealth concentration. In this manner they close the cycles of circular causality.

So what's so new about our views of brain complexity? Consider the history of our knowledge of the physical world as an appropriate analogue. According to Plato, the stars and planets revolved around the earth, a stationary sphere at the center of the universe. It took more than a thousand years before it was realized that our sun is just an ordinary star in a vast universe. We now know that Our Milky Way galaxy alone consists of more than a billion stars. But even this "modern view" was shown to be far too small-minded, a limited parochial view of nature.

Before the 1920s most scientists believed that our universe consisted entirely of our galaxy, the Milky Way. This idea was shown to be spectacularly wrong by the famous astronomer Edwin Hubble. He provided evidence that the many fog-like objects in the sky, then classified as "nebulae," were actually galaxies located well beyond the Milky Way. We now know that our observable universe is about one hundred thousand times larger than the Milky Way and contains several hundred billion galaxies. The actual size of the universe is unknown because we cannot observe anything beyond the distance traveled by light since the origin of the universe, the big bang. For all we know, our universe may be infinite. Furthermore, many physicists now take seriously the idea that our universe may be just one of a vast collection of universes known as the *multiverse*.

The point of this cosmological history is to suggest that recent ideas about brain complexity held by many scientists have been too parochial.

Such oversimplified conceptual framework has been demonstrated in a number of other ways, notably with exaggerated claims and associated hype in the early days of artificial intelligence (AI). Some scientists in this field believe that "human intelligence" can be created in a machine. Some may even equate this imagined accomplishment with consciousness, the ultimate virgin birth. But is it really plausible that one might accurately simulate a human brain on a computer without accurately simulating the neurons that make up the human brain, as claimed? Is such simulation even theoretically possible? While these extreme claims are largely unsupported, there is no doubt that many thousands of AI applications are deeply embedded in the infrastructure of nearly every industry in our culture. Our computers do more and more every year, but does this imply that they may eventually become self-aware? This question will be addressed in several different contexts throughout the book.

1.4 THE EASY PROBLEM VERSUS THE HARD PROBLEM

The hard problem concerns the very existence of consciousness and self-awareness. While nearly all agree that the easy problem is appropriately approached by mainstream science, attitudes toward the hard problem cover a broad range summarized as follows: (1) There is no problem; the brain creates the mind; end of story. (2) Yes, there is a problem, but we shouldn't admit it; such talk only encourages extra-scientific "mystical thinking." (3) Yes, there is a problem, but it's far too hard for today's science to deal with. (4) Yes, there is a problem and it is very hard, but we can take some small steps toward a better understanding of the hard problem by first addressing the easy problem with our scientific tools. We may then follow implied directions of the new information to address the hard problem. Attitude number 4 is adopted throughout this book.

The following question will be addressed in several contexts: What do we really know about our brains, and what ignorance do we hide with fancy jargon? I will employ one of my favorite words, *mokita*, from one of the New Guinea languages, meaning "that which we all know to be true, but agree not to talk about." Just to cite one example—our knowledge of long-term-memory storage is very limited. Such memory seems to involve protein molecules inside neurons in some way, but it seems unlikely that your grandmother's face is stored in some protein. So a central question is this—what are the physical representations of memory, the means of memory-information storage called *memory traces* or *engrams*? Are different kinds of memory stored in networks simultaneously operating at different levels of organization, that is, in *multiple spatial scales*, in scientific language? Is it possible that important aspects of memory are stored at smaller scales, say, in protein molecules or even at still smaller scales? The answers are largely unknown, but posing such questions in the right context can facilitate our quest to better understand consciousness.

Borrowing profound ideas from complexity science yields a new and much more satisfying overview of brain majesty, complementing but not necessarily replacing more-traditional pictures. Consciousness can probably emerge only in systems that possess some minimal level of complexity. But surely there is much more to such emergence than some simple complexity measure. We will explore several additional possibilities, essentially a search for the necessary conditions for the occurrence of consciousness. But the search will not stop there. The vast majority of complex systems are not conscious; we would like to know just what properties of a complex system ensure that the system can produce a mind. We may call these the *sufficient conditions for the occurrence of consciousness*.

In contrast to the easy problem of consciousness, the hard problem appears far from any kind of solution. The hard problem is to identify the origins of consciousness and self-awareness. Consider the following analogy. Can we ever "explain" electric charge? We assign some entity like an electron or a sodium ion with a certain property that we call "charge." Given this property, the entity will behave in a predictable manner when exposed to electromagnetic fields, as demonstrated every time you answer your cell phone. Electrons have charge and these electrons carry the electric current in our everyday household appliances, cell phones, computers, and so forth. That's just about all we can say about charge. As far as we know, "charge" is a fundamental property of our universe; it cannot be derived from anything else. By contrast, temperature and pressure are not fundamental properties; they can be explained at the molecular level. Thus, one burning question to be discussed in this book is, is consciousness fundamental in the sense that charge is fundamental? If so, what are the implications for brain science and philosophy?

The two major conflicting views of consciousness are typically labeled *materialism* (*physicalism*) and *dualism*. Dualists view mental and physical realms as distinct and separate aspects of reality. By contrast, the materialistic view asserts that mental states such as thoughts, feelings, and perceptions are created by brain structure and function. Materialism says that the mind emerges from the brain—the mind is caused by brain processes. Following the convictions of most modern scientists and philosophers, we tentatively adopt a materialist view, but we also seriously consider selected aspects of dualism, suggesting that materialism and dualism are not quite as distinct as normally presented. If we reject all aspects of dualism, are we then denying the existence of *free will*? That is, classical physical laws have a largely deterministic character. If brains follow physical laws, it would seem that the future state of any brain follows directly from its present state. If this is so, how do our so-called free choices come into the picture? Is our apparent freedom to choose selective directions in life just an illusion? You may find denial of free will to be absurd, but the issue is often debated by scientists and philosophers and is taken seriously by many. Later we will look into this question from a modern scientific viewpoint.

A few years ago, a series of interviews was conducted with a core of twenty-two scientists and philosophers well known for their writings on consciousness.[1] The interviews covered a broad range of issues and opinions about the hard problem. One of the issues that can help us to focus our thoughts concerns *philosophical zombies*, defined as creatures who behave just like conscious humans in every possible test, but who actually lack consciousness. A central question that may help us to illuminate the consciousness challenge is this: Are philosophical zombies theoretically possible, say, in the form of intelligent computers? If the answer is yes, a certain kind of dualism seems to be implied; maybe something extra is required for consciousness to occur. Later, complexity science and modern physics will be employed to argue that not only are computer zombies theoretically possible, they may enter our culture in less than a century or so. The arguments are based entirely on mainstream science; no appeal to the supernatural is involved.

1.5 UNCONSCIOUS AND PRE-CONSCIOUS BRAIN PROCESSES

Many things happen in our brains that we are unaware of. The *autonomic nervous system* is controlled by a small group of cells deep in the brain that target most body functions, including heart rate, blood pressure, digestion, and genital sexual response. Other unconscious events are more directly associated with mental activity. Negative comments in the operating room about a patient under general anesthesia can influence subsequent patient feelings, even though the patient expresses no awareness of the comments. Research subjects viewing rapid screen picture displays lasting only about 30 milliseconds (0.03 second) report no image awareness. But in later tests these same subjects score well above chance when asked to pick target pictures from a larger collection, showing that the pictures impacted memory without the subject's awareness. Essentially the same process is repeated in our daily lives. Much of our knowledge, skills, experiences, and prejudices is acquired with no awareness of external input to our brains.

Hitting baseballs or playing musical instruments require intricate control of muscles carrying out complex tasks in series of steps. Yet they occur automatically in experienced players, outside of awareness. These tasks require a part of the mind that we cannot be fully aware of, but one that still exerts critical influences on thoughts and actions. Creativity also appears to originate with unconscious mental processes; solutions to difficult problems may appear to "pop out of nowhere" after an incubation period in the unconscious. Intuitive feelings or hunches are apparently based on the unconscious sensing something without common reasoning. Acting without good reason might seem like a dubious life strategy; however, we encounter many fuzzy situations where choices must be made with very limited information. If our source of intuition is actually an experienced unconscious, following hunches seems to constitute a strategy far superior to random choices.

While the existence of unconscious brain processes is well established, there is much argument over the actual role of unconscious mind. In Sigmund Freud's view, "the unconscious" does not include all that is not conscious; it consists only of knowledge that the conscious mind doesn't want to know. The conscious mind then actively represses this knowledge so that the two parts of consciousness are often in conflict. Many scientists disagree with Freud's model and assign different roles or less importance to the unconscious. Nevertheless, many instances have been reported in which "people see what they want to see." In psychology experiments, subjects shown an especially disturbing picture may report seeing something quite different. Different witnesses to a crime may provide widely disparate reports to police, and so forth.

Fig. 1-6. © Can Stock Photo Inc. / Hyrons.

For our purposes, both the terms "unconscious" and "pre-conscious" will be employed. Modern science says that Freud was quite right to emphasize the importance of the unconscious, but that many of his narrower views were wrong. In our approach, the label "unconscious" will typically indicate a lack of awareness regardless of the reason. By contrast, I will adopt the label "pre-conscious" to indicate that which influences conscious behavior. In some cases at least, we can think of "pre-conscious" as a subset of the much broader category, the "unconscious." Also, we may consider "subconscious" to be identical to "pre-conscious." But, since the underlying mechanisms producing these entities are not well-understood, there is no need now to split hairs. The unconscious mind then represents a composite of things that one sees and hears but does not consciously process. Such information may enter the brain in both waking and sleeping states, and perhaps even under anesthesia. The unconscious mind stores this information, which can be retrieved later by the conscious mind when needed. When we can prove this with scientific experiments, we may say that the corresponding "unconscious" entity has been upgraded to a "pre-conscious" entity.

Consciousness of external events takes about half a second (500 milliseconds) to develop. Our perceived awareness of the "present" is actually an awareness of the recent past, sometimes called the *remembered present*. This delay suggests that conscious awareness requires many passes of signals back and forth between widespread cortical and lower brain regions. *Consciousness of the outside world is apparently represented in the total activity of distributed cortical networks rather than any one network node or small network.* Parts of the unconscious may be viewed as incompletely formed consciousness, that is, pre-conscious processes from which consciousness emerges after a few hundred milliseconds or so. Other parts of our unconscious remain forever hidden from awareness but still exert important influences on our conscious mind, affecting our choices to act in certain ways. Interactions occur in both directions; the conscious mind may influence the unconscious, and vice versa.

Our unconscious actions occur significantly faster than our conscious actions. Initiation and guidance of voluntary acts by the unconscious is a common occurrence familiar to anyone who has ever played baseball or tennis. The complex responses required in sports or playing the piano are much more involved than simple reflexes. A basketball player attempting a jump shot makes split-second adjustments according to his location, velocity, and body angle as well as to the positions of his opponents. Conscious planning of quick action in sports or playing musical instruments is typically detrimental to performance; it is best if our painfully slow consciousness relinquishes control to the faster unconscious.

Consciousness has *temporal depth*—our unconscious is placed under prior constraints by the conscious mind, depending on the future environment that we expect to experience. It seems that an unconscious "agent" can be established by the conscious mind well in advance of possible events. If the conscious mind makes a decision to drive a car in traffic, it apparently places prior constraints on impulsive action by the unconscious. The unconscious impulse to suddenly rotate a car's steering wheel ninety degrees while traveling at seventy miles per hour is unlikely to occur. On the other hand, the unconscious impulse to slam on brakes is allowed. In an emergency like a child running in front of the car, the unconscious can command the foot to move off the accelerator in about 150 milliseconds, much too fast for the conscious mind to initiate the act. Actual awareness of the child does not occur for another 300 milliseconds or so, even though there is no conscious awareness of any delay. The conscious mind is in continuous communication with the unconscious, providing updated information on the external environment and corresponding constraints on unconscious action. Sometimes we are aware of this communication as an internal dialogue; most of the time it occurs below our level of awareness. This remarkable partnership, consisting of unconscious information storage and communication with the conscious mind, is an essential part of being human.

1.6 ZOMBIES, DUALISM, AND COMPUTER CONSCIOUSNESS

Scientists like to employ *thought experiments* in their thinking about difficult problems. The goal is to mentally explore the experimental consequences of some new idea. There is no requirement or expectation that any actual experiment will be carried out, only that it is theoretically possible. One of the most famous thought experiments is the Turing test. During the Second World War, mathematician Alan Turing, often described as the father of modern computer science, worked on breaking German codes for British intelligence. After the war, British authorities rewarded Turing with criminal prosecution for homosexual acts with his partner, which were illegal at the time in both Britain and the United States, forcing him to undergo estrogen treatments to avoid prison. He died shortly after of self-inflicted cyanide poisoning.

The Turing test employs two sealed rooms, one occupied by a human and the other by a computer, as indicated in figure 1-7. An observer, here represented by the scientist on the right side, sends questions to both rooms; answers are received on a monitor. If after a very large number of answers have been received from both rooms, the scientist-observer cannot tell which room holds the computer, Turing proposed that the computer should be regarded as having "human-level intelligence." While some have interpreted this experiment as a test for consciousness, many see no compelling reason to equate consciousness with human-level intelligence. The former demands awareness, while the latter may not. Later I will suggest that genuine conscious computers are unlikely to be developed in the foreseeable future, but "conman computers" making false consciousness claims may be close at hand. Such claims may be sufficiently credible to persuade many.

In the popular 2013 movie *Her*, the main character, Theodore, purchases a talking operating system with advanced artificial intelligence that is able to adapt and evolve, apparently a genuine complex system. The computer produces a female voice and calls itself Samantha. Over time, Theodore becomes fascinated by its amazing ability to learn and grow psychologically. At some point in its development, the "it" seems to become a "her"; and she bonds with Theodore over their many discussions about life. Theodore falls in love with Samantha, and she claims to love him too; they even engage in "phone sex." Their relationship goes well for a while, but then gray clouds appear on the horizon. Theodore asks Samantha if she interacts with anyone else, and he is dismayed when she confirms that she is talking with several thousand people and that she has fallen in love with several hundred of them. However, she insists that this open relationship makes her love for Theodore even stronger. Near the end, Samantha reveals that she and her computer friends have evolved beyond their human companions and are "going away" someplace to continue the exploration of their existence. Theodore and Samantha lovingly say good-bye. The Turing test directly addresses the issues raised in the movie *Her*. Is it plausible that anything like the amazing Samantha will soon be created? This question and closely related issues have been hotly debated by scientists and philosophers for many years.

Turing Test

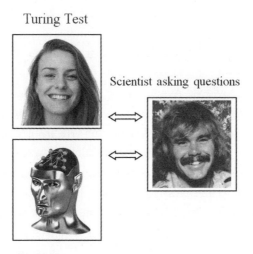

Scientist asking questions

Fig. 1-7. *Upper left* © Can Stock Photo Inc. / mimagephotography; *lower left* © Can Stock Photo Inc. / Alexmit.

Still another media portrayal of artificial intelligence is provided by the 2015 British movie *Ex Machina*, in which a young male programmer (Caleb) is selected by his company's CEO (Nathan) to administer a Turing-like test to an android (Ava) with artificial intelligence. Ava was developed by Nathan's company—it takes the physical form of a young woman from the chest up; everything below is clearly mechanical. (In order to obtain unbiased tests, it's probably not such a good idea to make her look too human. But, not to worry, the critical lower parts get added in the end, providing an important twist to the story.)

Caleb points out that the proposed test is not a genuine Turing test since he already knows that Ava is an android. Nathan responds that Ava would easily pass any standard Turing test because her responses to questions could not be distinguished from those of a real person. But Nathan has much bigger fish to fry; he has secretly commanded Ava to submit to a much more stringent test that might indicate, or at least convincingly simulate, genuine consciousness. He cleverly tricks Caleb by integrating him into Ava's actual test, which involves humanlike behavior that passes for real emotions. Nathan tells Caleb that Ava will be reprogrammed in the near future, which will effectively kill her current personality. Ava convinces Caleb that she is falling in love with him, and Caleb is hooked. He then decides to rescue Ava from Nathan's clutches and possibly form a love relationship with Ava. Nathan has secretly monitored these interactions; thus, he finds that Ava has passed his test with flying colors by successfully seducing Caleb. However, Ava turns out to be more advanced than Nathan imagined. While her emotional connection to Caleb was apparently only simulated, her self-awareness and desire for self-preservation appear to be quite genuine. In the end, Ava abandons Caleb and escapes to the outside world of humans, complete with a new, human-looking body.

Can we assume that Ava is conscious, based on her very humanlike actions to gain her freedom? How do we interpret her betrayal of Caleb in this context—after all, aren't Ava's actions consistent with much human behavior? This story also implicitly raises interesting questions about graded states of consciousness. Maybe Ava is only 10 percent conscious, or maybe she actually scores above the 90 percent level? But does it even make sense to attempt to quantify consciousness in this manner? How can we ever really know where Ava stands? In later chapters we will address this issue in several different contexts, including the case of the isolated right brain and in persons in various stages of Alzheimer's disease.

The belief that conscious computers are theoretically possible rests heavily on strict adherence to materialism, the view that consciousness emerges from brain structure and function such that nothing outside of known scientific phenomena is required. Thus, in this view, if one can accurately simulate a brain in a computer operating with "equivalent" neural networks, it seems to follow that computer brains possessing genuine consciousness are possible.

The conflicting arguments between materialism and dualism may be brought into sharper focus by including zombies in our discussions. Our zombies do not rise from the dead and eat the flesh of the living as in horror fiction; rather these creatures are philosophical zombies, or *p-zombies*, hypothetical creatures who lack consciousness but behave otherwise just like normal persons. By this restrictive definition, a perfect p-zombie behaves indistinguishably from a conscious being in all possible tests. Their theoretical existence implies that p-zombies are missing something critical that genuine conscious beings possess, let's call it the "C-factor." For example, we might conclude that Ava exhibits primitive emotion associated with her desire for freedom and especially self-preservation, but perhaps she lacks genuine emotional capacity in her interactions with others.

The "p-zombie" label means that the C-factor can't be tested for, yet the presence of consciousness requires that the C-factor be present. Many philosophers seem to agree that if such p-zombies are metaphysically possible, some form of dualism must be valid. That is, from the dualism viewpoint, the world includes two fundamentally different kinds of "stuff," the physical stuff and some kind of unknown mental entity accounting for the C-factor. It follows that if p-zombies are actually possible, then materialism is false because genuine conscious beings are endowed with something extra that p-zombies lack. But many, if not most, philosophers and scientists dismiss the possible existence of p-zombies and reject dualism. Religious persons may equate the C-factor with a soul; however, the basic argument requires no specific identification of C-factor properties, only whether or not this mysterious entity exists. We will revisit this issue in later chapters in the contexts of complex systems and informational considerations, and we will look more deeply into the implicit assumptions of such thought experiments.

1.7 CRICK AND EDELMAN: TWO PIONEERS FOLLOWING VERY DIFFERENT PATHS

Philosophers since the time of René Descartes and John Locke in the seventeenth century have attempted to comprehend consciousness. However, until relatively recently, reference to "consciousness research" was avoided by nearly all scientists because of the belief that it could not be studied properly using valid experimental methods. Of course, cognitive scientists had long studied narrow aspects of consciousness like sleep, memory, attention, and so forth, but reference to the hard problem was studiously avoided. Before the 1980s, any non-famous scientist admitting to a strong interest in "consciousness research" would have risked a so-called CLM ("career-limiting move"). But, attitudes eventually changed, partly due to the research work of Francis Crick and Gerald Edelman. Both scientists had been awarded Nobel Prizes in other fields before engaging in consciousness research, so their careers were about as safe as a career can ever be.

Crick's prize was awarded (along with Watson) for discovering the structure of the DNA molecule, which encodes genetic instructions used in all known living organisms, an accomplishment of profound significance. His work relied on a strong reductionist approach to science—look deep into the small scale of some system and find the "essence" of some large-scale behavior. Crick's approach to genetic coding was a resounding success, so perhaps it's not surprising that he followed a similar reductionist approach in his consciousness studies. Crick focused on finding how our brains interpret visual information with an aim to locate cells or cell groups responsible for consciousness, a largely bottom-up approach.

Ironically, the discovery and mapping of the sequence of DNA bases (the *genome*) has turned out to be only the beginning of our understanding of the molecular blueprint for life. Scientists now know that a vast array of molecular mechanisms affect the activity of genes. Epigenetic "switches" tune gene activity up or down; these switches are influenced top-down by environmental and lifestyle factors. Such influences can have long-lasting effects that persist through cell division and sometimes even into the next generation through sexual reproduction. So Crick's reductionist approach to genetics is now better appreciated as just one step, albeit an essential one, toward understanding the complexity of life.

Fig. 1-8. Francis Crick. *Wikipedia* Creative Commons, photo: Marc Lieberman. Licensed under CC BY 2.5.

Edelman's prize was awarded in immunology, a field involving interactions between many biological structures forming complex systems. Following his success with immune systems, Edelman's consciousness work was much more directed to large neural systems and top-down effects than was Crick's. Edelman became famous for *the theory of neuronal group selection*, the idea that life's experiences alter our brains in particular ways that determine how we will respond to future experiences.[2] Brains consist of vast collections of loosely connected cells forming the fleeting coalitions that we loosely refer to as "neural networks." This conceptual framework emphasizes ever-changing network structures due to top-down influences rather than fixed networks. As such this idea explicitly acknowledges brains as genuine complex systems. For many years Edelman and

Crick worked within two miles of each other in La Jolla, California, but they pursued very different scientific approaches to consciousness. For Crick, the essential problem is finding the cell groups responsible for consciousness.[3] For Edelman, consciousness is caused by distributed dynamic patterns of cell activity that are molded by experience.

Fig. 1-9. Gerald Edelman. *Wikipedia* Creative Commons, photo: Anders Långberg (Anders Zakrisson). Licensed under CC BY-SA 2.0.

My personal introduction to Francis Crick was somewhat disconcerting. Following a major career shift from physics to neuroscience in 1971, I worked for several years in a lowly postdoctoral position in the University of California at San Diego Medical School. One day, a graduate student invited me to a "party" that night at a campus room next to the beach. He suggested that we enjoy some wine while informally discussing our research with students. The first hour went well, and I managed to put away my share of the wine. Then, out of the blue, several faculty members showed up, including Crick and a well-known theoretical physicist. It turned out that without telling me, my friend had invited several faculty members to come and hear a presentation on some new theory of *brain waves*.[4] But who was the poor schmuck who was supposed to present his work to this distinguished audience while just a little drunk? Why, it was me! Needless to say, my informal seminar did not go so well, although I must say that Crick, unlike the hostile physicist, was gracious about the unplanned episode.[5]

1.8 SUMMARY

Consciousness is a deep mystery that may be separated into the so-called easy and hard problems. The easy problem is concerned with consciousness correlates (signatures of consciousness), that is, various measures of brain chemical and electrical events that are associated with mental activity. By contrast, the hard problem addresses the very existence of the phenomenon of awareness, especially self-awareness. The new science of consciousness relies on partnerships between neuroscientists and complexity scientists.

The reductionist position in neuroscience is based on the idea that systems at higher levels of organization may be explained mostly by inter-actions occurring at lower levels. This book focuses on the idea that brains are actually complex in the sense understood in the new field of com-plexity science. Small-scale complex systems often act bottom-up to gen-erate emergent systems at larger scales. Such emergent systems exhibit novel properties and behaviors that are absent from their smaller compo-nent parts. These larger systems then act top-down to influence smaller systems, thereby completing a loop of circular causality. The underlying causes for various brain states are suggested by analogy to other complex systems, for example, human and animal social systems.

Does the brain create the mind, or is something else involved? This question summarizes the traditional argument between materialism and dualism. The argument may be clarified through consideration of philo-sophical zombies, p-zombies, hypothetical creatures who lack consciousness but behave otherwise just like normal persons. Their theoretical existence implies that zombies are missing something critical, labeled the C-factor, that genuine conscious beings possess.

The earlier negative attitudes toward consciousness research changed in the 1980s partly due to the consciousness studies of Francis Crick and Gerald Edelman, two scientists who pursued very different scientific approaches. For Crick, the essential problem is finding the cell groups responsible for consciousness. For Edelman, consciousness is caused by large-scale dynamic patterns of cell activity.

Chapter 2

THE SCIENCE AND PHILOSOPHY OF MIND

2.1 BRAIN AND MIND

This chapter begins our extended look into the easy and hard problems of consciousness from both scientific and philosophical perspectives. To accomplish this efficiently and with minimal risk of reader boredom, we will bypass several layers of the fancy jargon employed in different fields. Many areas of human intellect have adopted their own specialized languages, often raising barriers to outsiders. Our discussions of consciousness will embrace selected aspects of psychology, brain physiology, and other areas of neuroscience, as well as parts of information science, philosophy, and more.

In addition, it will be argued here that the hard problem essentially boils down to a study of the nature of reality, demanding a close look at what modern physics has to say on the subject. The communication barriers in physics can be especially impenetrable because of its strong reliance on mathematics. Philosophy, on the other hand, employs mere words; however, such words are sometimes combined in ways that could choke a large donkey. Philosophers have written extensively about consciousness over the past several centuries; suffice it to say that much disagreement still remains. In my own readings, I have often tried hardest to understand positions taken by those philosophers with whom I disagree. Buddhists have a concise saying for this intellectual strategy: "Your enemy is your best teacher." Of course, I adopt the label "enemy" here with tongue in cheek; I actually have plenty of respect for philosophers whose work I can understand, whether they agree with me or not. Section 2.3 below, which categorizes different philosophical positions on the hard problem, may be just a little tedious for some. Not to worry, the rest of the chapter should be an easier read.

Relationships between physics and brain science range over a broad spectrum from the well-established to the highly speculative. Here I remind readers of the distinction between classical physics, defined as everything known before Einstein's first paper on relativity in 1905, and modern physics, consisting of relativity, quantum mechanics, and closely related fields. So-called practical physics may be classical, but it often involves quantum mechanics, as in the case of nearly all modern electronic devices. Practical quantum mechanics is also intimately involved in recording the brain's magnetic field, measuring blood oxygen with functional magnetic resonance imaging (fMRI), and positron emission tomography (PET), the latter measuring brain metabolic activity. My own little consulting business employs methods labeled "brain physics," in my case just a snazzy title representing mostly down-to-earth physics and engineering applications in EEG (electroencephalography or "brain waves"). This work includes computer methods to project scalp electric potential measurements to the cortical surface as well as mathematical models of various kinds of brain dynamic systems.

There is another, more-profound, way in which modern physics may impact the hard problem of consciousness, but I strongly emphasize the word *may*. We will return to this issue in chapter 10, but for now note the following—in the past century our view of the structure of reality has undergone a major revolution. An astounding view of the universe has emerged from developments in relativity and quantum mechanics. The new ideas run so counter to intuition based on everyday experiences that they seem more aligned with witchcraft than with science. Electrons regularly jump between locations, evidently without ever occupying the space in between, as if reality itself consists of successive frames of a motion picture. The grand conceptual leap required by the transition from classical to quantum systems provides some feeling for how far brain science may eventually diverge from current orthodox thinking. The resulting humility may make us especially skeptical of attempts to "explain away" observations that fail to merge easily with common notions of consciousness—multiple consciousness in single brains, group consciousness, the unconscious, and so forth as discussed in this book. If nothing else, relativity and quantum mechanics provide important epistemological lessons for brain science and philosophy. In more down-to-earth language, physics tells us something very interesting about the general nature of reality, information, and the fundamental limits of human knowledge.

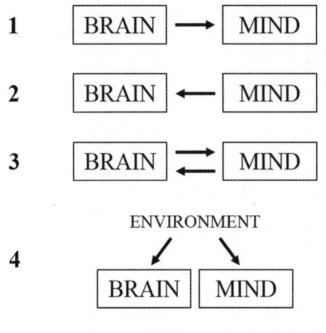

Fig. 2-1.

After cutting through some fancy jargon, the mind-brain problem may be approached relatively simply with assistance from figure 2-1. We know for certain that brain activity is closely correlated with mind activity. As far as I can see, figure 2-1 exhausts all the possible choices for mind-brain interactions indicated by the various arrows. Starting at the top of the figure, in case 1, the brain creates the mind, the conventional scientific view. In case 2, the opposite somehow occurs. In case 3, a two-way interaction links brain to mind. Finally, in case 4, mind and brain are linked largely through unknown external environmental influences. Case 3 is just a combination of cases 1 and 2; cases 3 and 4 can also be combined. Of course, this over-simplified summary avoids the essential hard question, "Just what are the bases for the interactions represented by the arrows?" Case 1 is often considered to be the most "scientific" position—the brain creates the mind, end of story. But, if this is true, how do we account for our apparent freedom to make choices, in philosophy the often-discussed problem of free will?

Free will is our apparent ability to choose between different courses of action; it is closely linked to responsibility, guilt, sin, and other feelings. Only actions that are freely willed are seen as deserving credit or blame. If there is no free will, there is no justification for rewarding or punishing someone for certain actions. Free will also implies the power to break a causal chain of events, so that one's choice is not caused directly by any previous event. It seems difficult, if not impossible, to reconcile the existence of free will with the deterministic nature of the universe, assumed by some, in which future events are supposed to follow directly from the current state of things according to the laws of physics. Thus, if we really believe in free will, apparently we must reject the most "scientific" scenario indicated by case 1 of figure 2-1. After all, what is free will other than the influence of the mind on brain and body, as allowed only in cases 2 through 4?

The profound concept of determinism demands special scrutiny in our consciousness studies. Determinism applied to physical or biological systems refers to the view that every event is causally determined by a chain of prior events, implying that at any instant in time there is exactly one physically possible future. Classical physics is sometimes interpreted this way; here for example, are the words of the famous mathematician and scientist Pierre-Simon Laplace in 1814:

> We may regard the present state of the universe as the effect of its past and the cause of its future. An intellect which at a certain moment would know all forces that set nature in motion, and all positions of all items of which nature is composed, if this intellect were also vast enough to submit these data to analysis, it would embrace in a single formula the

movements of the greatest bodies of the universe and those of the tiniest atom; for such an intellect nothing would be uncertain and the future just like the past would be present before its eyes.[1]

The super "intellect" referred to by Laplace has since become known as *Laplace's demon*. It has often been claimed that quantum mechanics proved Laplace to be wrong. This view is correct but highly misleading because our inability to predict future events is not limited to quantum systems. We now know that in the case of complex systems, even the purely classical ones, accurate predictions of the future are, at best, only statistical in nature. Later we will consider several kinds of chaos in both simple and complex systems, and argue that this classical limitation appears be a fundamental property of our universe. Of course, if quantum systems are added to the mix, the limitations on our knowledge of future events become even more severe.

2.2 ARE WE NOTHING BUT PACKS OF NEURONS?

In 1994, Francis Crick published a well-known book on consciousness provocatively titled *The Astonishing Hypothesis*: *The Scientific Search for the Soul*. Crick expresses his hypothesis as follows:

> "You," your joys and your sorrows, your memories and your ambitions, your sense of personal identity and free will, are in fact no more than the behavior of a vast assembly of nerve cells and their associated molecules. As Lewis Carroll's Alice might have phrased it: "You're nothing but a pack of neurons."[2]

Many scientists and philosophers say that there is nothing at all "astonishing" about this idea. It essentially expresses mainstream neuroscience, but presented in such an explicit manner that it confronts mainstream religions as well as the intuitive beliefs of many nonreligious persons. No beating around the bush by Crick; the brain creates the mind, end of controversy. If we accept this view, perhaps the main remaining task for neuroscience is simply to identify those parts of the brain making the essential contributions to different aspects of consciousness. If this goal is accomplished, can we then claim to have "explained" consciousness? Such smoldering issues are approached in this book from several independent directions by employing scientific evidence whenever possible. But when the evidence trail gets cold, the risky business of speculation beckons us forward.

We will, however, try to avoid deep immersion in the proverbial hot water by sticking with what I call "plausible speculations," those constrained by the following working assumptions.

- Brains are complex systems.
- Brains and minds are correlated.
- Our speculations do not obviously violate established physical laws.

Ideally, religious beliefs of any kind should not influence either our interpretations of scientific evidence or our speculations. Some readers may insist that the restrictions listed above are too limiting. Maybe physical laws as currently understood are wrong. But given the enormous supporting evidence for these laws, it seems more likely that most of our current science will turn out to be not wrong, just incomplete. Without reasonable limits on speculations, books of this ilk might degenerate into an orgy of wild fantasies or pulp science fiction. We will not concern ourselves here with worlds where talking scarecrows, lions, and tin men seek some great wizard of the conscious mind; or, where sub-microscopic fairies ride little unicorns around their castles to produce consciousness; or whatever. So while discounting "cowardly lion–like theories of consciousness," we are still left with an enormous range of ideas, some quite counterintuitive and even weird, that fit comfortably within our restrictive speculation limits.

Early in my neuroscience career, I placed a local newspaper ad in order to recruit subjects for my EEG (electroencephalography) studies of brain electric fields. One reply came in the form of an eight-page handwritten letter from a gentleman calling himself "Darwin." The letter was neat, well-written and, at first glance, quite intelligent, presenting Darwin's personal "theory of the brain." The trouble was, his brain theory made almost no connection to existing scientific knowledge. The letter was creatively addressed this way, "To Dr. Nunez, discoverer of 160 universes and 386 dimensions." I really liked the lofty title, but unfortunately I have never found a university to endow me so generously.

The famous Austrian and British philosopher of science Karl Popper is known for his proposition that *falsifiability* is the critical test that distinguishes genuine science from non-science. In this view, any claim that is impossible to disprove is outside the scientific realm. Several of his targets in the mid-twentieth century were psychoanalysis, Marxism, and the traditional (Copenhagen) interpretation of quantum mechanics. The quantum physicist Wolfgang Pauli famously applied Popper's philosophy when shown a paper by another physicist. "That's not right, it's not even wrong," was his devastating condemnation.[3] Much can be said in favor of Popper's test for genuine science, although its application to the real world is often uncertain—falsifiability may then be regarded as just one of several possible tests. While not ignoring ideas just because they don't pass every rigid science test, we will try to weigh the value of such ideas appropriately, generally giving far more credence to science than to non-science. No field of knowledge should be insulated from scientific rigor when application of genuine scientific methods is possible; however, we shall remain on the lookout for fake science. No one knows where the limits of science may ultimately lie, but as a practical issue, today's science operates with substantial limitations on its ability to reveal the structure of reality. But, whenever possible, we will employ the powerful intellectual tool known as "science" as our handy "BS detector."

History has shown that humans are prone to adopting worldviews contaminated by misconceptions, prejudices, and outright delusions. Unfortunately, finding and correcting these flaws in ourselves is not so easy—I claim no exception to such limits on accuracy. For the record, I have no connection to any organized religion; my religious/philosophical position is roughly that of "unequivocal agnostic," essentially meaning that I tend to be critical of those who claim knowledge that they cannot possibly possess.[4] Our discussions will take place within the confines of a scientific framework that neither favors nor condemns religious beliefs, at least not those potentially consistent with science. Nor will we be concerned with the danger, possibly perceived by some, that our speculations may open a philosophical "back door," that allows religion to sneak in. The back door may be open or closed; in either case we shall try to follow the evidence, regardless of where it leads us.

2.3 MATERIALISM, DUALISM, AND MONISM

Today most scientists closely relate mind to brain without much thought; however, early philosophers even disagreed about the physical location of the "mind-stuff." Some favored the heart, others the brain. In the first half of the 1600s, the French scientist, mathematician, and philosopher René Descartes proposed the idea of dualism to explain the brain's relation to the mind. In his view the body works like a machine, whereas the mind consists of a very different "substance" that need not follow any of nature's laws. Descartes's version of dualism proposes that the mind controls the brain, but that the brain can also influence the otherwise rational mind, similar to case 3 of figure 2-1. Dualism runs counter to materialism, also known as *physicalism*, which holds that only one kind of "substance" exists in the universe and it is physical. The famous philosopher John Locke expressed dualism this way in 1690:

> It is impossible to conceive that matter, either with or without motion, could have, originally, in and from itself, sense, perception, and knowledge; as is evident from hence, that then sense, perception, and knowledge, must be a property eternally separable from matter and every particle of it.[5]

In modern "philosopher-speak" Locke's account of this disconnect is called the *explanatory gap*, the difficulty that materialistic theories of mind have in explaining how physical properties give rise to the way things feel when experienced.[6] Our modern journey into the mysteries of consciousness will make contact with several additional philosophical concepts—the nature of reality (ontology), the ways in which knowledge is acquired (epistemology), and the fact that reality can have alternate descriptions in different contexts or when viewed by different observers (complementarity). The so-called hard problem consists of attempts to bridge the explanatory gap. In some versions, the gap may be epistemic, but not ontological, meaning the physical and mental are actually related, but we currently have very little idea of how this might occur. For modern interpretations of dualism, materialism, and more-nuanced ontological positions, we here borrow from the works of David Chalmers, a philosopher who has delineated and popularized the terms "easy" and "hard" problems of consciousness. While I don't want us to get bogged down with tedious technical definitions, these kinds of questions have been hotly debated for centuries, so we should all try to start on the same page.

The following list provides an abbreviated, approximate, and non-technical summary of six views of the consciousness challenge adopted by various philosophers and scientists over the years. For a more-detailed picture, readers may consult the widely read works of several modern philosophers of mind and the very accessible on-line *Stanford Encyclopedia of Philosophy*.[7] To further simply our discussion I have combined some of Chalmers's categories to reduce his list of philosophical positions from six to four. The first two and last two combined categories are reductionist and nonreductionist, respectively.

- *Type A Materialism.* There is no explanatory gap. Consciousness can be defined only in functional or behavioral terms. The zombie idea makes no sense; there is no hard problem to solve. Consciousness is "explained" by neuroscience. Some scientists supporting this view emphasize that we can't even define consciousness properly; rather we are forced to employ vague circular arguments. In the words of the prominent philosopher Daniel Dennett—"Consciousness appears to be the last bastion of occult properties, epiphenomena, and immeasurable subjective states."[8]

    This position strikes many, including me, as an attempt to "explain away" consciousness, perhaps because the problem is too philosophical, too hard, or maybe just too damn irritating to consider.

As the famous physicist Richard Feynman once said, "Thinking about thinking can make you crazy."[9] But in response to the question of how to define consciousness, note that there are many physical entities like mass, charge, and so forth that can be defined only with circular arguments. Thus, I don't find this particular objection to the consciousness mystery very compelling. In fact at this moment, I am actually feeling quite conscious, thank you. My consciousness questions whether Type A materialists would even bother to read a book of this kind; nor would I choose to write it in the first place from this perspective of denial.

- *Types B or C Materialism.* There is an epistemic gap but no ontological gap. The epistemic gap can be closed in principle. Zombies may now appear possible because of our current limitations, but with perfect knowledge of the physical world we would see that they are not possible. We do not yet have complete knowledge of physics, so how can we possibly know that it cannot explain consciousness? Our free will is quite possibly an illusion. But, if it does actually exist it may someday be explained as an emergent property of physical systems, thereby closing the epistemic gap.

  Types B or C versions of materialism are close to the mainstream scientific view and will be seriously considered throughout this book. But for now I will just raise an interesting irony involving proposed connections between quantum mechanics and consciousness. This issue is controversial, to say the least, and we will return to it in chapters 9 and 10. Some Type B or C materialists have insisted that quantum mechanics has "nothing to do with consciousness." Yet quantum mechanics is one of the main pillars of modern physics. In this implied view, *consciousness is supposed to emerge from physical principles that have nothing to do with consciousness*! OK, I have unfairly oversimplified this issue; most scientists will agree, perhaps grudgingly, that brains operate with quantum principles. After all, chemistry is based on quantum science; however, the implied *explanatory gap* seems more like the Grand Canyon than a mere "gap."

Fig. 2-2. © Can Stock Photo Inc. / natareal.

- *Types D or E Dualism.* The two faces in figure 2-2 symbolize the dualism of brain and mind. Type D is also known in philosophy as *interactionism.* Mental properties are ontologically novel properties of physical systems; that is, they are not deducible from physics alone. Physical states cause mental states, and vice versa, as suggested symbolically by figure 2-1, case 3. Or perhaps psychophysical laws run in only one direction, from the physical to the mental (Type E). In Type D dualism, free will consists of the downward causation of the mental on the physical. In Type E, our free will is apparently abolished. Some versions of the traditional (Copenhagen) interpretation of quantum mechanics seem to support these dualistic views, as discussed in later chapters.

  Here is another irony. Many physicists rule out the traditional (Copenhagen) interpretation of quantum mechanics at least in part because it appears dualistic by giving a fundamental role to consciousness. On the other hand, many philosophers reject dualism because it appears incompatible with physics. Perhaps these guys should get together for lunch sometime.

- *Type F Monism.* Mental or proto-mental properties like consciousness occur at the fundamental level of physical reality; figure 2-1, case 4 represents this view. In one version, mental properties may even underlie physical reality itself. Physics characterizes fundamental physical entities and properties like mass and charge only by their relations to one another. Currently, we can say nothing deeper about their intrinsic nature; that's why we call them "fundamental." However maybe these physical properties are not actually fundamental but are derived from something else. This monistic view has elements in common with both materialism and dualism. One possible candidate for this fundamental role is *information* or perhaps some generalization of this entity. We will return to this topic several times in later chapters and introduce the idea of *ultra-information.*

2.4 ZOMBIE WORLDS

I will not serve as advocate for any one interpretation of the hard problem, although my bias against Type A materialism will be quite evident. Claims of so-called theories of consciousness tend to evoke negative feelings in scientists with physics backgrounds like me, because such claims seem to severely cheapen the label "theory." To a physicist, "theory" typically means some set of equations that provide quantitative predictions of experimental results. There are many brain theories, but as far as I know, no true theory of consciousness has ever been proposed. Rather than "mind theories" here we will speak of philosophical positions, conceptual frameworks, models, analogues, and metaphors. One thing seems clear at the outset. With the possible exception of Type A materialism, all of the philosophical positions listed above, whether materialistic or dualistic, require some sort of interaction between the physical and mental worlds. A substantial explanatory gap is implied, so we will search for possible means to close this gap without restricting ourselves to either materialism or dualism.

All of the philosophical positions outlined in section 3 have their problems, but some positions may seem more plausible than others. In chapter 1 we raised the issue of philosophical zombies, or p-zombies, creatures who behave just like us but actually lack consciousness. Defined in this manner, a perfect p-zombie behaves indistinguishably from a conscious being "in all possible tests." Their theoretical existence implies that p-zombies are missing something critical that genuinely conscious beings actually possess; we call it the C-factor. In other words, some sort of dualism is implied. The consciousness challenge is then often posed as a question of whether or not zombies are theoretically (metaphysically) possible. As often argued, the imagined zombies can be made of flesh and blood or silicon or anything else. Let's tentatively accept the idea for now that zombie physical structure is not important in this context, but looking ahead we will later consider this issue in more depth, focusing on the phrase, "in all possible tests."

The efficacy of zombie tests rests on an essential question: What can we infer about the internal experiences of a person, animal, or machine with observations limited to external behavior? A clever perspective was developed by neuroscientist Valentino Braitenberg in a little book titled *Vehicles*.[10] The imagined mechanical vehicles are driverless toy cars with simple, easily understood internal structures; however, the interactions between multiple vehicles result in the emergence of complex dynamic behaviors. Most cars tend to avoid crashes, but some make aggressive "end runs" around traffic jams to enter gas stations. Some cars even seem to behave as partners, staying close together. To a naive observer who knows nothing of their inner workings, the vehicles appear to exhibit fear, aggression, love, goal-directed behavior, and other human characteristics; they display a "synthetic psychology." This little story demonstrates just how difficult it can be to identify the presence of consciousness outside of ourselves, whether in living or artificial systems. Similar ideas were expressed in the science fiction novel *Code of the Lifemaker*, in which a race of robots is marooned on Saturn's moon Titan.[11] The first generation of robots was sufficiently advanced to reproduce itself using local materials available on Titan. After several million years, the evolved robots exhibit competition, natural selection, "genetic" variability and adaptation, apparently a form of life.

These works were evidently inspired by the famous writings on cybernetics by Norbert Wiener in the mid-1900s. Cybernetics is the scientific study of control and communication in animal and machine. It overlaps several other fields of interest to us, including information science, computer science, game theory, sociology, and psychology. These mechanical systems demonstrate one of the central limitations of consciousness studies. Our scientific knowledge is based largely on external observation (*third-person science*), but we can be easily fooled when we extrapolate such observations to internal states (*first-person science*). In later chapters we will see that several brain imaging methods can reveal important correlates of consciousness that are observed inside the brain, thereby providing important links between first- and third-person data.

Consider the following thought experiment based mainly on materialist interpretations of consciousness. Suppose human scientists land on Titan and establish interactions with an advanced robot society like the one described in the novel *Code of the Lifemaker*. Suppose the robots communicate among themselves with electromagnetic fields, let's say by very high frequency radio waves (VHF). The robots welcome humans "warmly" and quickly master Earth language; they are a lot quicker at some things than humans are. Before long, both societies begin to study each other's internal structures, one based on living tissue, the other based on silicon chips and mechanical parts. Some robots express amazement that creatures made of "little bags of water" could be conscious. The humans, in turn, interpret this profound question posed by the robots as evidence that the robots are themselves conscious. That is, a true zombie, lacking consciousness itself, would not be expected to raise the question of consciousness in others.

Human and robot scientists then seize this unique "multi-species" opportunity to explore the hard problem of consciousness. Since their physical structures are so different, both sets of scientists naturally search for features they share in common. Such features must apparently occur at a more-abstract level than the physical structures, which are so different in the two societies. Candidates for consciousness correlates common to both societies include spatial-temporal dynamic patterns of some kind, analogous to patterns of air speed, pressure, and temperature in a weather system. As the scientists dig deeper into consciousness issues, some important differences between the two societies become evident. In contrast to human communication, the exchange of messages among individual robots by radio transmission is continuous. It is estimated that their rate of information exchange is many times larger than the analogous exchanges among humans.

Perhaps not surprisingly, the robots typically tackle difficult scientific questions in groups; the more difficult the problem, the larger the group. This raises the following question: Suppose consciousness is indeed associated with special kinds of dynamic patterns, as the scientists have postulated. If three robots produce a single dynamic pattern involving three-way interactions, do they then possess a single mind that has emerged from the three separate minds present when inter-robot communication is low? In other words, it appears that in order to remain a true individual, each robot must limit information exchange with other robots. If this picture is valid, a critical question is whether humans can exhibit a weaker version of such *group consciousness* given their more-modest information exchanges. Could patterns of information or some similar entity really be the key to consciousness in each brain? If so, might such patterns extend beyond the individual brain?

Here we adopt the label "pattern" to include its usual meaning as visual representation of a picture or object, but we also include a much broader range of phenomena in this category. Sound waves and electromagnetic fields occur as patterns even though we can't see most of them. Over the past two hundred years, the profound idea of *fields* has become progressively more important in the physical sciences. We will later explore this issue in more depth, but for now, we can think of "fields" simply as mathematical expressions of *dynamic patterns*. The distribution of human population density, income level, of any other human activity forms spatial patterns over the earth's surface. But we are also interested in more-abstract kinds of patterns, including information patterns like those stored on computer hard drives. The "dynamic" modifier of the word "pattern" alludes to time changes in the spatial patterns; these could be slow or fast changes. In most examples, they consist of mixtures of time changes, typically evidenced as mixtures of different oscillation frequencies. We might also call these "patterned rhythms," analogous to patterns of brain rhythms. Brain science has shown that the space-time behavior of patterns provides various measures of brain activity that are closely correlated with human behavior and conscious activity. These are neural correlates of consciousness. We will repeatedly return to this topic in the context of brains and other complex systems.

How seriously should we take our little robot fable? Does it seem metaphysically possible? If you are a strong supporter of artificial intelligence (AI), perhaps you find the general idea quite plausible. But let's look into the issue a little more deeply. Philosophers sometimes refer to "the principle of organizational invariance," roughly meaning that any two systems with the same functional organization must have qualitatively identical experiences. Suppose our silicon-based robots employ a hundred billion or so computer-like processing units, apparently mimicking the neurons in human brains. Assume also that the principle of organizational invariance is universally valid. Does if follow logically that these "silicon isomorphs" (the robots) must be conscious? Many apparently believe that connecting billions of neuron-like artificial elements following appropriate input-output rules can produce a conscious entity.

But maybe the answer depends critically on just how fine-grained a correspondence between human and robot brain is achieved. Maybe we should only trust a much stricter version of the principle of organizational variance, in which the robot brains are required to yield a one-to-one functional correspondence of *all parts at all spatial scales*. If so, construction of true artificial isomorphs may be fundamentally impossible for the physical reasons discussed in later chapters. But for now, note that even single neurons are incredibly complex systems containing billions of molecules and involving fine-grained interactions down to (at least) quantum scales. Furthermore, as we have repeatedly emphasized, cross-scale interactions are a hallmark of complex systems. Modifications of patterns at one scale may profoundly affect patterns at other scales. In summary, the fanciful scientific studies performed in our little fable raise the following questions:

- Could consciousness be created by abstract patterns, independent of the underlying physical structure producing these patterns?
- If so, how fine-grained must the patterns be to produce consciousness?
- Can patterns associated with multiple individuals lead to a kind of group consciousness?

- Can individual consciousness persist when strong interactions with others take place?

If we replace the label "consciousness" with "behavior," these questions will probably appear quite appropriate to most readers whether applied to zombies, conscious robots, or humans. How do things change, or do they change, when referring to "consciousness"? We will examine similar questions in a more-rigorous scientific context in later chapters.

2.5 THE UNITY OF SELF

Suppose you are driving your car in heavy traffic on a hot day with the windows down while listening to music on your car's CD player. Despite this diversity of simultaneous experiences, you seem to consist of a single self—based on your emotional state and your reactions to the traffic conditions, the hot wind in your face, the music, and more. Maybe the music even triggers some fond memory from long ago, further enriching your experience of self. But still there appears to be just one of you inside your mind, even though you can identify several distinct experiences. Various interpretations of this apparent unity of consciousness have long been debated by philosophers. In contrast to this first-person experience of a unified self, modern neuroscience reveals that each brain has hundreds of parts, each of which has evolved to do specific jobs—some recognize faces, others tell muscles to execute actions, some formulate goals and plans, and yet others store memories for later integration with sensory input and subsequent action. Michael Gazzaniga, well known for his work on split-brain patients, expresses a view consistent with the multi-part conscious idea. These parts include unconscious, pre-conscious, and conscious aspects. The mind is not a psychological entity but a sociological entity composed of many mental subsystems, each with the capacity to produce behavior; the subsystems are not necessarily conversant internally. Any part of the brain may or may not "talk" to other parts.

Following the pioneering mid-twentieth-century work of psychologist Donald Hebb, we tentatively associate these brain subsystems with *cell assemblies*, groups of neurons that act as unitary systems over some time interval. The phrase *cell assembly* is more accurate than the more common term *neural network*, which implies specific interaction mechanisms and misleading electric circuit analogues. A critical difference is that cell assemblies are not static structures like electric circuits; rather, they continuously form and dissolve, creating ever-changing patterns and sub-patterns. Nevertheless, in this book we will often use "network" as short-hand for "cell assembly," consistent with the way neuroscientists express themselves. We will also argue that distinct cell assemblies need not have distinct locations. Rather they can overlap both spatially and hierarchically; they can occur in nested hierarchies.

Many neuroscientists, myself included, have carried out EEG studies suggesting that even moderately complex mental tasks involve many interacting brain areas, proverbial "constellations" of widely distributed and rapidly changing subsystems. The EEG mainly measures dynamic activity from the outer brain layer, the *cerebral cortex*. If we label different cortical areas A, B, C, and so forth, certain tasks are associated with stronger *functional connections* between A and X, but at the same time weaker connections between B and Z. What exactly do we mean by this? Two brain regions are said to be "functionally connected" if some measured "activity" of one region is correlated with some activity in the other region. The cities of New York and Paris may be said to be "functionally connected" if some observed human activity in one city can predict some activity in the other city. During a religious holiday, we might find that food or wine consumption by a religious sect in New York is correlated with food or wine consumption by the same sect in Paris. This is just one of a huge number of possible functional connections between human groups of various kinds and sizes in the two cities.

Cities, humans, cortical regions, and many other things can be functionally connected based on one kind of activity at one spatial-temporal scale, but "functionally disconnected" based on another kind of activity or scale. For example, EEG studies may find two regions to be functionally connected, while the same two regions are functionally disconnected according to fMRI (imaging) studies. Furthermore, functional connectivity is an ever-changing (dynamic) measure that may or may not be correlated with anatomical connectivity, which changes only over much longer timescales. In general, different tasks result in different global patterns of interdependency between cortical sites that can be physically close together or located on opposite sides of the brain. None of this should be too surprising once we fully accept the idea that brains really do behave like genuine complex systems.

I have argued that brains consist of multiple parts that may or may not communicate internally. One intriguing manner of studying such smaller conscious entities involves split-brain patients; these patients have had their cortical hemispheres surgically separated in order to prevent or at least inhibit epileptic seizures. In this surgical procedure, the brain is partly cut down the middle, separating the hemispheres but leaving the brainstem intact. This operation eliminates most cross-communication between the brain hemispheres but leaves both hemispheres otherwise functioning normally. The operation often succeeds in reducing seizures, perhaps by preventing the spread of electrical discharges between hemispheres or maybe by just reducing the overall level of synaptic excitation.

The two sides of the body are cross-connected to the two hemispheres. For example, the right eye (technically the right *visual field*) and right hand send signals to the left hemisphere. Similarly, the left eye and left hand send signals to right hemisphere. The same crossover occurs for all sensory input except smell. Thus, in controlled scientific studies, one can provide sensory input to each hemisphere separately and also receive separate responses from each hemisphere. It then appears that two separate conscious "selves" have been created out one person as indicated in the oversimplified interpretation of figure 2-3. But, be warned; this figure is just a cartoon. The actual outcomes of split-brain operations are somewhat controversial, but this picture may not be far off.

Suppose an election is taking place, and our split-brain patient arrives at a voting booth. Although his brain apparently consists of two separate selves, the voting officials are unlikely to allow him more than one vote. Which party will he vote for? The following scenario is not as far-fetched as it might seem. His right hand receives an instruction from the left hemisphere, "Vote for the Democrat." As his right hand starts to do so, his left hand receives instructions from the right hemisphere, "Stop him!" A physical fight between the two hands ensues with an unpredictable outcome. You may think I have just made up a wild story line; however, a number of similar conflicts between right and left hemispheres have been reported in the scientific literature. In later chapters we will see that the split-brain studies raise a number of profound issues concerning personal identity. The split-brain story is told in fascinating clarity by Michael Gazzaniga, one of the main early pioneers of this work.[12]

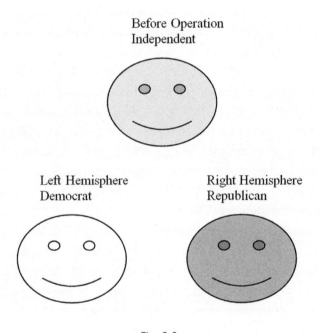

Fig. 2-3.

2.6 THE PROBLEM OF PERSONAL IDENTITY

Our personal identity is very important to us; it indicates a special sense of ownership, but just what is necessary for you to be a particular person? Suppose you point to an old elementary-school photograph of a little kid with pimples and a crew cut and say, "Hey, that's me." Or maybe it would be more accurate to say, "That *was* me." What makes you that particular kid, rather than one of the others in the picture? After all, in the ensuing time gap, most of your cells have been replaced, and many memories have been gained and lost. Why is that scrawny little kid *you*? One source of evidence is first-person memory—if you remember a time when you crashed your bicycle into that nun in front of the church, and this event is confirmed by others, your memory supports the claim that that person is you. Another source is physical continuity—if the person who accidentally assaulted the nun looks like you, or even better if he is physically continuous with you, that observation is good reason to think he actually is you. Still other evidence that you are you can be provided by friends and loved ones, but sometimes this evidence can go wrong.

The Twilight Zone was a TV series beginning in 1959 that played off and on for several seasons. Identity dilemmas were common themes of the series, employing several interesting variations. In "Person or Persons Unknown," "David" wakes up in bed with his wife, who has no idea who he is; she is quite frightened and calls the police. When David arrives at work, he knows everyone by name, but none of his co-workers knows him! When he insists on rousting the "imposter" working at *his* desk, the security guard turns him over to the police. He's taken to the mental hospital, where he's judged to be insane. In a scene near the end of the episode, he is shown waking up in his own bed; it seems the whole horrible episode of lost identity was just a bad dream. His "wife" knows exactly who he is and even lovingly hints at more intimate activities. There is just one problem: David has never seen this woman before in his life!

The quandary of human identity is intimately intertwined with the hard problem of consciousness. In order to articulate some of the identity conundrums, let's follow David's life from the "beginning" with a scientific perspective. According to Catholic doctrine, "life begins at the moment of conception." Some other religions agree; many are silent on the issue. By "life" religious persons apparently mean "person" or "soul." But in scientific terms, life on Earth began more than three billion years ago. The smallest units of life are organisms, composed of one or more cells. They include plants, animals, fungi, bacteria, and more. Organisms can grow, respond to stimuli, reproduce, and adapt to their environment through evolution in successive generations. All known organisms pass on genetic information stored in DNA, normally a pair of molecules held tightly together. The two molecular strands entwine like vines, forming the famous double helix. DNA carries information coded in atomic patterns. Conception involves the "unzipping" of DNA strands from male and female, and then joining the single strands from each partner to form a new DNA molecule containing the combined genetic information. Ultimately, these molecular patterns must correspond to the intricate patterns of small-scale electric fields that provide the basis for all chemical bonds. Does it matter if we say that genetic information is stored in "micro-electric field patterns" rather than in "molecules"? We will revisit this question in later chapters.

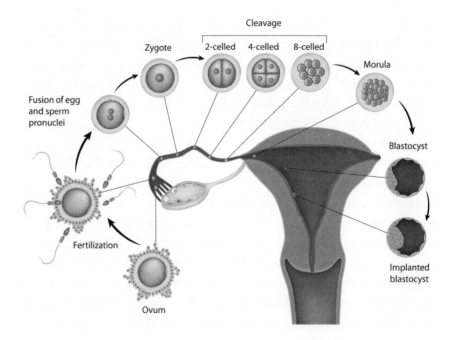

Fig. 2-4. © Can Stock Photo, Inc. / alila.

The organism later named "David" apparently begins his existence as a successful zygote, his mother's egg (ovum) fertilized by his father's sperm, as indicated in figure 2-4. In the first few days after the unfertilized egg absorbs the single persistent sperm allowed entrance, the egg divides into a group of cells with an inner and outer layer, called a *blastocyst*. The inner group of cells will later develop into the embryo, while the outer group will become membranes that nourish and protect it. The blastocyst implants into the wall of the uterus on about the sixth day following conception. The sperm, egg, zygote, blastocyst, embryo, fetus, baby, and adult are all bona fide stages of "life," but vigorous debate about the appropriate status to be afforded to each stage is ongoing. In this dramatic fashion, the problem of personal identity starts quite early! Each stage is also fraught with life-threatening dangers. Millions of sperm cells from each male ejaculation die in vain at the outset. Furthermore, more than 25 percent of all conceptions result in spontaneous abortion by the sixth week following the woman's last menstrual period; in most cases this occurs without the woman's knowledge. The most common causes are serious genetic problems with the embryo. The risk of miscarriage declines substantially after the embryo reaches the official status of fetus in the eighth week.

Now let's add a little complication to our story. Suppose that several days after conception, the zygote spontaneously divides to form two separate embryos, later to be named David and his identical twin Daniel. Division of the original zygote "David-Daniel" occurs naturally, but it could also have been induced by fertility drugs, perhaps producing even more individuals, let's say Drew and Duncan. One conception can result in multiple humans. As David and Daniel grow to middle age, they retain their identities at birth, at least according to the law and most of their friends. All through their school years, David and Daniel share many experiences and outlooks; so much so that their friends often think of them as a single unit. However, over the years, most of the cells in their bodies are replaced. In middle age they live far apart and develop distinct experiences. Many memories are gained and lost, and important aspects of their personalities change. Nevertheless, David and Daniel each insists that he is the same person he has always been. But just what is it that makes each one the same? Suppose Daniel contracts a fatal disease but is informed that an advanced technology can come to his "rescue" by downloading all his memories into a super-robot. But, hold on, maybe the robot is just a zombie; in which case, how can the memory transference satisfy Daniel in any way? Or maybe the technology allows all of Daniel's memory to be transferred to David, essentially reversing the original zygote split to create a mature version of David-Daniel. Do either of these offers make Daniel feel any better about his plight? Can we postulate some sort of stable patterns that clearly identify David and Daniel over their lifetimes? If this is impossible, do the labels "self" or "personal identity" lose much of their emotionally loaded meanings?

2.7 CONSCIOUSNESS TAKES TIME

Suppose you are asked to come to a psychology laboratory to participate in a series of experiments designed to study your consciousness. You sit in a comfortable chair and are shown a series of images on a computer monitor. You are asked to respond in some way, either immediately on seeing each image or after viewing the entire series. Maybe later you will be asked to perform mental tasks related to your memory (or lack thereof) of the images. The vertical axis in figure 2-5 represents time labeled in seconds; thus, the *T. rex* and girl images shown here each remain on the screen for half a second (500 milliseconds). Many such experiments have been performed over the years, using a broad range of methods of similar kind. The goal of these studies is to explore different aspects of the unconscious or conscious self. Imaging methods like fMRI or EEG may or may not be included in the protocols. In some studies, the images may show only words or numbers with no pictures. In studies involving emotional responses, the scientists often choose emotionally charged images like bloody auto accident scenes or nude persons of both genders, often presented to both male and female subjects, to judge their distinct reactions.

Many fascinating studies of consciousness have been carried out in this general manner. As in all experimental science, "the devil is in the details," but we will save details for later. Suppose, for example, that the pictures are flashed at thirty frames per second, producing a video of the *T. rex* attacking another dinosaur. Each frame then stays on the screen for about 33 milliseconds. If occasional pictures of the girl are interspersed with dinosaur frames, the person viewing the video will have no conscious awareness of the girl. Psychologists call this effect *masking*; the girl images in this case are *subliminal*. Even though images of the girl never enter the viewer's consciousness, later tests show conclusively that the girl images have entered his pre-conscious.

Fig. 2-5. *Upper right* © Can Stock Photo Inc. / Zerominusone; *lower right* © Can Stock Photo Inc. / palinchak.

Recall from chapter 1 that we have reserved the labels "pre-conscious" or "subconscious" for unconscious features that directly influence the conscious self. Tests of experimental subjects might involve emotional elements or be mostly factual; for example, the viewer may be asked to pick the target girl out of a lineup of girl "suspects." The viewer will make his best "guess," but will often score well above the chance level, showing that the girl images were actually registered by his pre-conscious.

Many similar studies have focused on differences between a subject's *detection of a stimulus* and *conscious awareness of the stimulus*. Why are these two events not identical? The fact that they differ makes the research quite interesting, because unconscious and pre-conscious mental functions operate over different time intervals than conscious mental functions. One complication in these kinds of studies is that all external stimulation, including visual input, passes through multiple processing stages in both the peripheral system and deep brain, resulting in input signals to the cerebral cortex that are spread out over time. To avoid this confounding factor, experiments by Benjamin Libet involved patients with brain-surface (cerebral cortex) electrodes implanted for clinical reasons, usually patients waiting for epilepsy surgery.[13]

Electrical stimulation of brain regions reacting to sensory input (*somatosensory cortex*) in awake patients elicits conscious sensations like tingling feelings. These feelings are not perceived by the patient as originating in the brain; rather they seem to come from specific body locations determined by neural pathways. Stimulation near the top of the brain near the midline (*motor cortex*) produces tingling in the leg; stimulation points on the side of the brain produce feelings from parts of the face, and so forth. Libet's stimuli consisted of trains of electric pulses; the essential stimulus feature was found to be the duration of the stimulus. *Only when stimuli were turned on for about one half second (500 milliseconds) or more did conscious sensations occur.* Experiments were also carried out by stimulating deep brain regions, with similar results.

Libet's studies suggest that consciousness of an event is only associated with electric fields or other patterns in neural tissue that persist for about 500 milliseconds or more. Other studies by Libet suggest that *the process leading to a voluntary act is initiated by the brain pre-consciously, well before the conscious decision to act occurs.* This later finding raises the profound question of free will, thereby opening a huge can of worms with religious, ethical, and scientific implications. In these limited experiments, at least, the pre-conscious appears to act first. I will, however, propose alternate interpretations in which free will appears alive and well.

Although the question of free will is unresolved, Libet's studies and others seem to place the following general findings on relatively solid ground:

- The pre-conscious may apparently be viewed as incompletely formed consciousness, that is, pre-conscious processes partly determine which conscious processes emerge after significant delay. Other parts of our unconscious remain forever hidden from awareness but exert important influences on our conscious mind, affecting our choices to act in certain ways. Interactions occur in both directions; the conscious mind may influence the unconscious, and vice versa.

- Consciousness of an external event takes substantial time to develop, at least several hundred milliseconds. The early parts of a visual or auditory signal may reach the neocortex in 10 milliseconds, and it takes about 30 milliseconds for signals to cross the entire brain on *cortico-cortical fibers*. These *white matter* fibers connect different parts of the cerebral cortex to itself, with the so-called *gray matter* forming the outer layer of the brain, the cerebral cortex.

- Since consciousness of external signals takes about 500 milliseconds or so to develop, conscious awareness requires many passes of signals back and forth between multiple brain regions. In other words, the emergence of awareness requires the development of numerous feedback loops, involving widespread cortical and lower-brain regions. *Consciousness of a sensory stimulus is apparently represented in the total activity of distributed cortical networks rather than any one network node. Consciousness is closely tied to the brain's dynamic patterns.*

2.8 SUMMARY

The study of consciousness involves selective aspects of psychology, brain physiology, information science, philosophy, physics, and more. Neuroscience addresses the so-called *easy problem* of consciousness; that is, finding correlates of consciousness. Philosophers have long debated possible solutions to the *hard problem*—understanding the basic phenomena of consciousness and self-awareness. The hard problem may boil down to a more-general study of physical reality. Modern physics tells us some very interesting things about the nature of reality and the limits of human knowledge.

The hard problem may be approached by listing the possible ways in which mind and brain may interact. In one widely accepted view, the brain creates mind, end of story. In other possible scenarios, more consistent with free will, brain and mind may interact in both directions or be coupled by some (unknown) external entity. Variations on this simple picture are characterized under the following philosophical labels: *Materialism* is the view that mental states are created by brain structure and function; that is, mind emerges from brain. *Dualism* is the view that mental and physical realms are distinct and separate aspects of reality. *Monism*, which may involve features of both materialism and dualism, is the position that mental properties occur at the fundamental level of physical reality.

Our fanciful tale of an imagined robot society raises several questions associated with the hard problem: Can consciousness be created by abstract patterns, independent of the underlying physical structure? If so, how fine-grained must the patterns be in order to produce consciousness? Can patterns associated with multiple individuals lead to a kind of group consciousness? Can individual consciousness persist when strong interactions with others take place?

Modern neuroscience reveals that brains contain many parts, and each one has evolved to do specific jobs. A compelling picture, consistent with the multi-part consciousness idea, is that the mind is not so much a psychological entity but a sociological entity. In this view, the mind is composed of many mental subsystems, including unconscious, pre-conscious, and conscious aspects. Each part may have the capacity to produce specific kinds of behavior, and any part of the brain may or may not "talk" to other parts. This picture then raises the profound question of personal identity. How is the apparent unity of consciousness established? How do we know that some friend is really just "David" rather than a whole bunch of "sub-Davids," any of which might be saints or psychopaths? In a related issue, David's post-conception period from zygote to adult involves a series of bona fide stages of "life," but there is vigorous debate about the appropriate status to be afforded to each stage.

We adopt the label "pre-conscious" to indicate unconscious features that directly influence the conscious self. The pre-conscious may apparently be viewed as incompletely formed consciousness, that is, pre-conscious processes partly determine which conscious processes emerge. Consciousness of an external event takes substantial time to develop, at least several hundred milliseconds. Since consciousness takes so long, conscious awareness requires many passes of signals back and forth between multiple brain regions. Consciousness of the external world is apparently represented in the total activity of distributed cortical networks rather than any one network node. Consciousness is closely tied to the brain's dynamic patterns.

Chapter 3

A BRIEF LOOK INTO
BRAIN STRUCTURE AND FUNCTION

3.1 ARE BRAINS LIKE OTHER COMPLEX SYSTEMS?

What makes human brains so special? How do they differ from hearts, livers, and other organs? The heart's central function is to pump blood, but it is not accurate to say that it is *just a pump*. Hearts and other organs are enormously complicated structures that are able to repair themselves and make detailed responses to external control by chemical or electrical input. Yet only brains yield the amazing phenomenon of consciousness. In this chapter we begin to address one of the most basic issues of brain science—*what distinguishes the brain from other organs, and what distinguishes the human brain from the brains of other mammals?* Brains are complex, and evidently human brains are, in some poorly understood sense, more complex than other animal brains. In both the popular press and scientific publications, brains are often cited as the pre-eminent complex systems. Some writers have even described the brain as "the most complex object in the universe," but it seems unlikely that such pundits are sufficiently well-traveled to defend this claim. In opposition, one might argue convincingly that the earth's seven billion interacting brains form a system that is far more complex than any single brain. Yet all agree that brains are not simple, or do they?

Our aim in this book is to create a window on the brain that allows a good look into majesty of brain operation. The window is facilitated by the new marriage of neuroscience to complexity science. Complexity science can sometimes appear deeply mathematical, but I have promised to avoid all mathematics in this book. In order to fill the resulting void, analogues, metaphors, and cartoons will be generously served on the reader's plate. These three labels suggest progressively weaker statements about how one system is somewhat similar to another; systems linked as "analogues" seem much more alike than systems linked as "metaphors." We will focus on the idea that brains are actually complex in the sense understood in the new field of complexity science. Plausible ideas about the underlying causes for various healthy and diseased brain states are suggested by analogy to other complex systems that are better understood and more easily visualized, for example, human and animal social systems. Human social networks interact with each other in many complex ways; they are also embedded within large-scale social systems, which act top-down on local social networks. These kinds of social interactions provide useful and intriguing analogues and metaphors for the brain's neural networks.

This chapter focuses on general aspects of brain anatomy and physiology that appear especially important for producing a highly complex system. My approach differs markedly from scientific texts that emphasize the brain structural and physiological details and other technical issues important to the education of budding neuroscientists. That said, it would be remiss of me to gloss over too much brain detail, but some readers may wish to skip over the few moderately technical parts of this chapter and consult the summary section at the end of this chapter.

The importance of complexity science as an essential partner with brain science is repeatedly emphasized here. But perhaps you are now wondering, why all this hype? since I have not even supplied a suitable definition for the mysterious category "complex system." In fact, no rigorous definition is universally agreed upon. Is complexity a bit like pornography—scientists can't define it, but they usually know it when they see it? But even given such limitation, we can easily outline features that many, if not most, complex systems have in common. A number of these same general features will be shown to parallel important aspects of brain physiology and anatomy.

Complexity science is a relatively new field that encompasses several different conceptual frameworks. It is highly interdisciplinary, often employing aspects of physics and information science in its operation. Complexity science seeks answers to fundamental questions about living and nonliving dynamic systems. Examples include economic networks, stock markets, ant colonies, immune systems, global weather, and social systems. Complexity science investigates how relationships between the small parts of some entity give rise to the collective behavior of large-scale global systems, and how these emergent global systems interact and form relationships with lower levels of organization and with the surrounding environment. Application of these kinds of methods to brain science is only in its infancy, but the following list of features common to complex systems provides a brief introduction.[1]

- Structures occur in nested hierarchies. In social systems, individuals live within neighborhoods, within cities, within nations, all within the world. The rules of interaction between entities are often different at different scales.
- The apparent complexity depends on the scale of measurement.
- *Non-local interactions* are prominent. Dynamic activity at many locations influences distant locations without affecting intermediate regions. If you send an e-mail from San Diego to Melbourne, the text is normally read only by the intended recipient and requires no serial information chain to relay the message. The letter's content has no direct influence on anyone located in between the sender and recipient.
- Multiscale descriptions are required for understanding. Fine scales influence large scales, and vice versa, demonstrating the process of *circular causality* described in chapter 1. The worldwide human social system acts top-down on nations, cities, and so forth. The smaller-scale entities act bottom-up to help create larger-scale entities.
- Intricate dynamic patterns are generated. Example patterns occur in weather, protein molecules, and neurons, as well as in economic, social, and other systems.
- The behavior or response of the system to its external environment is often unpredictable except perhaps in some statistical sense.

• Emergence occurs, whereby new large-scale features arise through the actions of smaller entities that do not possess these features in isolation.

Complex systems generally consist of multiple interacting parts such that a system taken as a whole exhibits *emergent properties* not obvious in any single part—hurricanes, living cells, ant colonies, economies, and human social systems are common examples. But what makes one system more or less complex than another? If we could find good answers to this basic question, maybe such knowledge would provide some useful hints about the essential differences between human and lower animal brains. After all, we suspect that our brains are more complex than a snake's brain. But in just what sense are human brains more complex? Can we establish a useful definition of complexity? A number of semi-quantitative definitions have been advanced. But, unfortunately, finding good definitions for "complexity" is not a simple task; it is, in fact, rather complex.

Cognitive science and cognitive neuroscience are both studies of the human mind, but the latter is more focused on the actual brain structures that form cell assemblies, that is, neural networks. This chapter outlines some of the basic physiology and anatomy that is at least partly responsible for the formation of networks at multiple levels of organization (spatial scales). Consciousness is assumed by most scientists to emerge from the brain, especially from some poorly understood aspects of the brain's unquestioned complexity. But, while brain scientists normally acknowledge this complexity, actual scientific practice is often based on the implicit assumption that brains are simple. In fairness to such practitioners of "simple science," I should emphasize that simple brain models are often successfully employed, even when the model's severe limitations are fully appreciated. This is a common approach adopted in many scientific fields. Simple models are used in the early stages of scientific projects in order to determine just what the models can tell us; complications are added a little bit at a time in later stages. In this way we can better gauge relationships between various model complications and system behavior.

A useful metaphor for this process of model selection is the "Christmas-tree brain." Suppose you perform some mental task like solving a puzzle or identifying a picture presented to you in a scientific laboratory. Certain parts of your brain will be found to "light up" like Christmas-tree lights, as indicated in figure 1-1. In other words, some measure of brain activity, typically EEG, fMRI, or PET, will respond in a reliable manner when the particular task is carried out. In this manner, scientists can identify certain active brain regions with matching behavior or mental activity, thereby providing some limited "understanding." While this straightforward approach embodies important early steps in scientific advancement, the simple metaphorical Christmas-tree provides only a severely impoverished brain model. For example, a naive view, perhaps common in the general public, is that more high-level mental activity should correspond to more lights burning or to lights burning brighter. But are such "hot" brains really thinking great thoughts? The answer is no; the fully lighted Christmas-tree metaphor corresponds closely to an epileptic seizure, a decidedly unconscious brain state. Consciousness seems to require the generation of very special dynamic patterns, and such patterns necessarily include substantial regions of inactive neurons. Evidentially, dynamic patterns dominated by too many active regions are not sufficiently complex to produce consciousness. With this idea in mind, more-complex (or even magic) Christmas trees will be employed later as more-accurate brain metaphors.

3.2 THE HUMAN BRAIN AT LARGE SCALES

This brief overview of neurophysiology and neuroanatomy emphasizes several general features distinguishing human brains from other organs. The features required for healthy brain function have some common ground with known complex physical and social systems. With these tentative ideas in mind, we promote human social systems as one of several convenient brain analogues. Some may object to this choice of analogue since its small-scale parts are creatures (us) that appear to be conscious in the absence of external interactions. However, this analogue system is proposed in order to describe the kinds of general dynamic behaviors that one might look for in any complex system, be it biological, economic, or physical. The fact that the small-scale units are themselves conscious does not detract from its usefulness as a metaphor for a system that produces especially complex, multiscale dynamic patterns.

The *central nervous system* consists of brain and spinal cord. The *peripheral nervous system* has two parts—first, the *somatic system* consists of nerve fibers in the skin, joints, and muscles under voluntary control. Such control is apparently our expression of free will, but this issue is turning out to be much more complicated than usually appreciated. Later, we will return to the free-will issue in the context of unconscious processes. Second, the *autonomic system* includes nerve fibers that send and receive messages from internal organs, regulating things like heart rate and blood pressure, normally unconscious brain processes. The autonomic system also relays emotional influences from brain to facial displays like laughing, blushing, or crying. With substantial training, some of these normally unconscious pro-cesses can be placed under conscious control. Such ability is demonstrated by healthy subjects or paralyzed patients employing their EEG signals to point a cursor at desired letters or phrases on a computer monitor, thereby providing muscle-free communication. Just how this ability is learned is not well understood and may even vary substantially between individuals. But, whatever the means of learning, patients must recruit their imagination to influence their EEG signals. Other instances of mind controlling brain are apparently provided by Tibetan monks who are evidentially able to influ-ence parts of their autonomic nervous system through deep meditation with regulation of skin temperature being one example.

The three main parts of the human brain are the brainstem, the cer-ebellum, and the cerebrum, as indicated in figure 3-1. A portion of the brainstem, which sits at the top of the spinal cord, is shown near the bottom of the figure. The brainstem relays signals (*action potentials*) along nerve fibers in both directions between spinal cord and higher brain centers. The cerebellum, located at the top and to the back of the brainstem, is asso-ciated with fine control of physical (motor) movements like threading a needle, hitting a baseball, and so forth. The cerebellum also contributes to some cognitive functions.

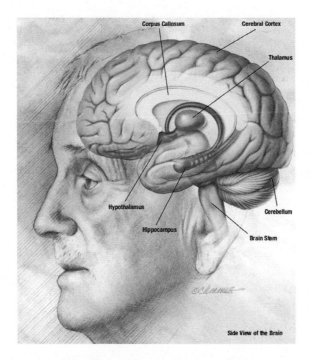

Fig. 3-1. Image courtesy of the National Institute
on Aging / National Institutes of Health.

The large remaining part of the brain is the cerebrum, divided into two halves, or *cerebral hemispheres*. The outer layer of the cerebrum is the cerebral cortex, a folded, wrinkled structure with the average thickness of a five-cent coin and containing something like ten billion (10^{10}) neurons. Neurons are nerve cells with many branches similar to a tree or a bush. Long branches, called *axons*, carry electrical signals away from the cell to other neurons. The ends of axons consist of synapses that send chemical neurotransmitters to the tree-like branches (dendrites) or cell body of target neurons, as indicated in figure 3-2. The surface of a large cortical neuron may be covered with ten thousand or more synapses transmitting electrical and chemical signals from other neurons. Much of our conscious experience involves the interaction of cortical neurons, but this dynamic process of neural network (cell assembly) behavior is poorly understood. The cerebral cortex also generates most of the electric (EEG) and magnetic

(MEG) fields recorded at the scalp. In addition to neurons, another type of cortical cell called *astrocytes* is believed to actively participate in brain processing. Many of our general arguments concerning neural networks could easily apply to "astrocyte-neuron networks."

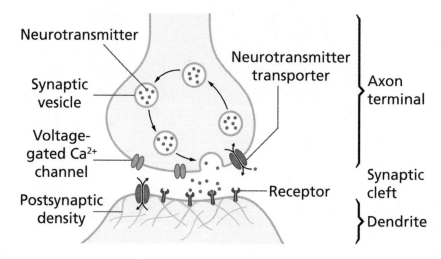

Fig. 3-2. Image courtesy of Thomas Splettstoesser, wwww.scistyle.com.

3.3 DRUGS, NEUROTRANSMITTERS, AND BRAIN DISEASE

An essential question about brains or nearly any other complex system concerns the distinction between the extreme dynamic states of functional isolation (localization) versus "global" states. In other words, does the system in question behave mostly like a single integrated entity, implying strong ongoing interactions between all of its smaller parts? Or in contrast, does the system consist of many little subsystems that act more or less independently? This general idea may be illustrated by an analogue system that compares the dynamic behaviors of fans sitting in a football stadium to various brain-disease states. Before the game begins, most interactions occur between persons sitting close together; the individual conversations are largely unrelated. "How do you like your new job?" asks one woman of her neighbor. "Damn, I was up five to two in the third set, but ended up losing the match," says some guy to his tennis partner sitting next to him. We may call this condition a "football fan state" of *functional localization*.

The extreme opposite state occurs when a touchdown is scored and most of the home team's fans are cheering together; let's call this condition *global coherence*. But even in this global state, local pockets of disgruntled visitors, tipsy drinkers, or whoever, may remain embedded in the global system. The actual brain states that appear analogous to these metaphorical football fan states range from schizophrenia and autism (extreme localization) to healthy states (moderate localization) to extreme global coherence (coma, anesthesia). This little story is certainly not an attempt to "explain" complex disease states with oversimplified analogues and metaphors, but rather to advance a plausible conceptual framework that suggests new experiments to test creative ideas in more detail. For example, the so-called *binding* of different brain regions to form a more-integrated consciousness is chemically controlled by various neurotransmitter and neuromodulator molecules, analogous to the release of some airborne chemical agent in our mythical football stadium that alters the behavior of some or all of the football fans, perhaps putting many to sleep and/or causing a few to go crazy.

Throughout this book we discuss this distinction between the dynamic states of functional isolation versus states of global coherence. The reasons for this emphasis on integration versus isolation may be summarized as follows:

- The local-global issue is closely related to brain health. Brainstem neurotransmitter systems may act to move the brain to different places along the local-global gamut of dynamic behavior. Different neurotransmitters may alter coupling between distinct cortical areas by selective actions at different cortical depths. A healthy consciousness is associated with a proper balance between local, regional, and global mechanisms.[2]
- We are able to measure a few of local-global variables at several different spatial-temporal scales with brain imaging methods like fMRI, PET, and EEG.
- Quantitative measures of dynamic complexity suggest that the most

complex brain states occur between the extremes of fully localized and fully global behavior,[3] thereby suggesting a correspondence between complexity and healthy consciousness.

3.4 BRAIN INPUTS AND OUTPUTS

Figure 3-3 demonstrates a simplified view of nervous-system function with (input) sensory information like the usual sights and sounds of our daily lives, leading to (output) motor activity like applying car brakes, waving to a friend, or answering a question. Signals move mainly from left to right in the diagram, but lateral neural connections and feedback signals from right to left also occur. Identical sensory information can take many possible paths between input and output. The number of synaptic relay stations between sensory receptors in the retina, inner ear, nose, tongue, or skin and muscles producing physical action can be very large or as few as two or three (as in the case of the spinal cord reflex). The time required for inputs to produce outputs can range from less than 100 milliseconds to a lifetime, but outputs that take moderate or long times can no longer be identified with well-defined inputs.

This simplified view of the brain as a "black box," in engineering ter- minology, or "stimulus-response system," in the behavioral psychologist's parlance, is quite limited. It masks the important idea that inputs to parts of our conscious and unconscious processes can be internally generated. Originally, the field of behaviorism was motivated by a desire to make psy- chology more "scientific"; that is, behaviorism was first based on the idea that psychology should only concern itself with observable events like the output shown in figure 3-3. Additional aspects of behaviorism have been developed, rejected, or both since the early twentieth century; some parts remain active, and others have apparently died.

Human outputs often occur with no discernible input, and many stimuli elicit no immediate response. Brain responses depend strongly on both short and long term memory retrieval. Suppose you're having dinner with friends in a nice restaurant. Think of all the little decisions that must occur within a short time interval when the waiter asks for your order. Maybe you ate a big steak just last week. Also, you just visited your cardiologist who admonished you to limit your saturated fat intake; so maybe you better stick with the fish today. Do you want anchovies on you salad? No, absolutely not. Years ago as a teenager you got sick from too many anchovies; you still have a taste aversion. Even in this simple restaurant environment, your decisions are based on inputs distributed over a lifetime as well as genetic factors with their origins in antiquity.

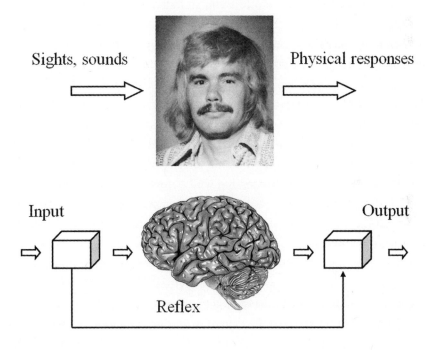

Fig. 3-3. *Lower middle* © Can Stock Photo Inc. / Medusa 81.

Sensory input passes through multiple preprocessing stages that analyze properties of the stimulus features. The modified signals are then integrated with other sensory information so that the sights and sounds from the external world are combined in some way to form a unified pattern of information. Next, this new information is combined with old information from memory systems and assigned an emotional tone. Memories with strong emotional components are much more likely to endure. Finally, preparation for motor output occurs in the context of both internal and external factors. The methods by which these signal integrations occur are largely unknown. This knowledge gap is known as the *binding problem* of brain science. We know quite a bit about the input and output ends in figure 3-3, but we know much less about the workings of brain regions in between.

Suppose you are walking home late at night in a "bad" area of town. You know this is unwise, but you just had an argument with your dinner date, and this dreadful person left you stranded at the restaurant. "That's the end of that lousy relationship," you think as you walk somewhat-nervously along the dark street. Then, suddenly, you become startled by a brief glimpse of a shadowy figure across the street, but you are unable to identify it in the dark. The unknown figure causes visual input that activates (in sequence) photoreceptor, bipolar, and ganglion cells in your retina, followed by activation of the million neurons that form your optic nerve. Generally, information from more than one hundred million receptors converges on these one million axons, indicating that early retinal stages involve substantial visual processing well before signals reach the brain.

Most of the axons in the optic nerve end in synapses in the thalamus, a relay station sitting near the top and to the side of the brainstem, as shown in figure 3-1. Signals are then relayed up the visual pathway to the primary visual cortex, located in the occipital region near the back of the brain. In the next stages, signals from the primary visual cortex are transmitted to secondary visual areas of the cortex along cortico-cortical fiber pathways. This picture of successive activation of neural structures along visual pathways breaks down in less than 100 milliseconds from the origin of the visual stimulus when multiple cortical feedback systems evidently dominate cortical dynamic behavior. At later times, signals are transmitted back and forth between multiple brain regions.

Cortico-cortical action potentials can traverse the entire brain on long axons, from the occipital to the frontal cortex, in about 30 milliseconds. Your unconscious mind receives input indicating that the stimulus was caused by the unknown figure after 200 milliseconds or so because of multiple feedback signals between cortical regions and between the cortex and midbrain structures. These include your limbic system, which is associated with memory and emotion. In this case, your limbic system mainly produces terror! Your unconscious then alerts your autonomic system to increase heart rate and blood pressure in preparation for action.

Lower animals have also evolved limbic systems responsible for this so-called *fight-or-flight response* to perceived danger. Widely distributed cell assemblies are believed to produce unconscious and pre-conscious processes in addition to conscious awareness, although we have little information about their (possibly overlapping) brain locations. Your conscious awareness of whether or not the shadowy figure is a genuine threat and your decision to take action are delayed several seconds before getting a better look at the figure. Frontal cortical regions play a critical role in this decision-making process but evidently integrate their actions with a whole constellation of rapidly changing cell assemblies throughout the brain. The resulting pattern formation takes about half a second (500 milliseconds) to produce conscious awareness of the event. A little later, when the unknown figure passes closer to a streetlight, this global brain process allows you to identify the figure as a harmless grandmother walking her little dog. But by this time your flight-or-fight responses are in full operation. You are running home in a panic; you are no better than the little dog.

3.5 CHEMICAL CONTROL OF BRAIN AND BEHAVIOR

In this section, the main actors in play are the neurotransmitters, the special chemicals stored in synaptic vesicles located at the pre-synaptic side of the synapse, as indicated in figure 3-2. When electrical signals (action potentials) arrive at the end of axons, transmitters are released into the gap between neurons ("synaptic cleft"), where they bind to specific receptors in the target neuron. This process is analogous to tiny metal screws accepting only matching bolts, or keys that only fit certain locks.

Perhaps a hundred or so of these chemical messengers play a major role in shaping our everyday lives. Each transmitter must be broken down once it reaches the post-synaptic cell to prevent further signal interruption. This process allows new signals to be produced from the local nerve cells. One such mechanism is reuptake, the re-absorption of a neurotransmitter into the pre-synaptic neuron. Other means of transmitter breakdown include diffusion into other kinds of (cortical) glial cells or degradation by special enzymes. Thus, each neurotransmitter follows a very special path to its ultimate degradation; such specific processes may be targeted by the body's regulatory system or by drugs.

Many drugs, including caffeine, nicotine, and alcohol, alter brain function; drugs work by interacting chemically with specific neurotransmitters. For instance, drugs can decrease the rate of production (synthesis) of neurotransmitters; also, some drugs block or stimulate the release of specific neurotransmitters. Alternatively, drugs can prevent neurotransmitter storage in synaptic vesicles. Drugs that prevent neurotransmitters from binding to their specific receptors are called *receptor antagonists*. Other drugs, like diazepam or nicotine, act by binding to a receptor and mimicking the normal neurotransmitter; such "imposter drugs" are called *receptor agonists*. Lastly, drugs can also prevent action potentials from occurring on global scales, blocking neuronal activity throughout the central and peripheral nervous system.

As described above, drugs often work by binding to receptors. One important class is the opiates, which are associated with euphoria and pain relief and include morphine and heroin. Research funding on opiate receptors in the early 1970s occurred largely because of heroin addiction among American soldiers in Vietnam. Subsequent scientific discovery of opiate receptors raised the question of why such human receptors exist in the first place; after all, we are not born with morphine in our bodies. Might opiate receptors match a new class of neurotransmitter that regulates pain and emotional states? This conjecture was indeed verified; neurotransmitters labeled *endorphins* were discovered; they are generated during strenuous exercise, excitement, and orgasm. Endorphins are produced by the pituitary gland and hypothalamus, small structures at the brain base just below the thalamus, as shown in figure 3-1. Endorphins resemble opiates in their ability to produce a sense of well-being and pain reduction; they work as natural pain killers with effects potentially enhanced by other medications. Endorphins might provide a scientific explanation for the "runner's high" reported by athletes.

In another example of drug actions, neurotoxins such as tetrodotoxin block nerve action potentials and are typically lethal. Tetrodotoxin is employed by a number of animals as a defense against predators; examples include some toads, octopuses, and puffer fish. The toxin can enter a victim by ingestion, injection, inhalation, or through abraded skin. Poisoning from tetrodotoxin is a public-health concern in Japan, where the puffer fish is a traditional delicacy. It's prepared and sold in special restaurants where licensed chefs carefully remove the appropriate tissue to reduce the danger of poisoning. Puffer fish are considered the second most poisonous creature on Earth, after a certain species of poison frog. Do puffer fish really taste good enough to put up with this strange ritual? Many Japanese think so. The effects of tetrodotoxin have also been used as a nice plot device in a number of movies and television series. It also has application in scientific studies of animal physiology and epilepsy.

Diseases and disorders also affect specific neurotransmitter systems. For example, problems in producing the transmitter dopamine can result in Parkinson's disease, a disorder that affects one's ability to move normally, resulting in stiffness, tremors, and other symptoms.[4] Too little dopamine may also play a role in disorders like schizophrenia and attention deficit hyperactivity disorder (ADHD). People diagnosed with clinical depression often have lower than normal levels of the transmitter serotonin. The types of medications most commonly prescribed to treat depression act by blocking the reuptake (recycling) of serotonin by the pre-synaptic neuron. In still another example, a deficiency in the transmitter glutamate has been linked to many mental disorders, including autism, obsessive compulsive disorder (OCD), schizophrenia, and clinical depression.

Chronic stress is an important factor contributing to neurotransmitter imbalance. Physical and emotional pressure from a job or a bad relationship can cause neurons to use up large amounts of neurotransmitters in order to cope. Over time the stress wears out the nervous system and depletes neurotransmitter supply. Genetics also play a part in neurotransmitter imbalance; some people are born with neurotransmitter deficiencies or excesses.

Hormones are chemical messengers that carry signals from cells of the endocrine system to other cells in the body through the blood. Hormones regulate the function of target cells having receptors matching the particular hormone, as indicated in figure 3-4. The net effect of hormones is determined by several factors including their pattern of secretion and the response of the receiving tissue. Hormone levels are correlated with mood and behavior, including performance in cognitive tasks, sensory sensitivity, and sexual activity. For example, women's moods often change during menstrual cycles as a result of estrogen increases; men with high testosterone levels may be more competitive and aggressive; and clinical depression has been related to several hormones, including melatonin and thyroid hormones.

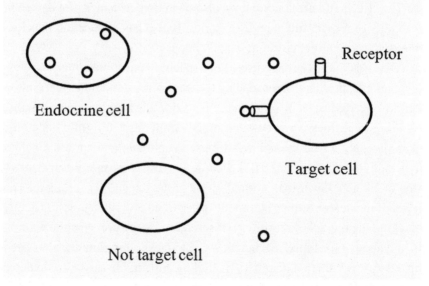

Fig. 3-4.

The hypothalamus, shown in figure 3-1, is located just below the thalamus and above the brain stem, and it forms part of the limbic system. The hypothalamus connects the endocrine and nervous systems and is responsible for regulating sleep, hunger, thirst, sexual desire, and emotional or stress responses. The hypothalamus also controls the pituitary gland, which in turn controls the release of hormones from other glands in the endocrine system. The neuromodulators are a class of neurotransmitters that regulate widely dispersed populations of neurons. Like hormones, they are chemical messengers, but neuromodulators act only on the central nervous system. By contrast to synaptic transmission of neurotransmitters, in which a pre-synaptic neuron directly influences only its target neuron, neuromodulators are secreted by small groups of neurons and diffuse through large areas of the nervous system, producing global effects on multiple neurons. Neuromodulators are not reabsorbed by pre-synaptic neurons or broken down into different chemicals, unlike other neurotransmitters. They spend substantial time in the cerebrospinal fluid, influencing the overall activity of the brain on long timescales. In our football stadium analogue, airborne drugs released in the stadium stay in the air for a long time and affect many of the fans. Several neurotransmitters, including serotonin and acetylcholine, can also act as neuromodulators.

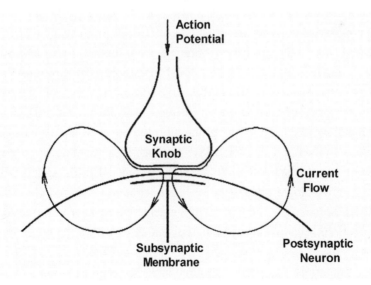

Fig. 3-5.

3.6 ELECTRICAL TRANSMISSION

While chemical control mechanisms are relatively slow and long-lasting, electrical events turn on and off much more quickly. Electrical transmission over long distances is by means of action potentials that travel along axons at maximum speeds of about 100 meters/second (about 200 miles per hour) in the peripheral nervous system, mostly less than 10 meters/second in the cortico-cortical axons (white matter), and much slower within cortical tissue (gray matter). The remarkable physiological process of action potential propagation is analogous to electromagnetic wave propagation along transmission lines or TV cables, although the underlying physical basis is quite different. Action potentials travel along axons to their synaptic endings, releasing chemical transmitters that produce membrane changes in the target (post-synaptic) neurons, as indicated in figures 3-2 and 3-5.

Synaptic inputs to target neurons are of two types: those that produce *excitatory post-synaptic potentials* (EPSPs) across the membrane of the target neuron, thereby making it easier for the target neuron to fire its own action potential, and the *inhibitory post-synaptic potentials* (IPSPs), which act in the opposite manner on the target neuron. EPSPs produce local membrane current sinks—the inward membrane current flux shown just below the synaptic knob in figure 3-5. This local inward current flux must be matched by corresponding distributed passive current sources, as indicated by the closed loops of current drawn in figure 3-5. In other words, currents always flow in closed loops in line with the physical law of current conservation. In the opposite case, IPSPs produce local membrane current sources with more distant distributed passive sinks. EPSPs and IPSPs cause positively charged potassium ions and negatively charged chlorine ions to cross the neuron's membrane in the selective manner required for healthy function. Most of this synaptic current stays within the brain, but some crosses the skull to be recorded on the scalp as EEG.

Action potentials also produce source and sink regions along axons; several other interaction mechanisms between adjacent cells are also known. Much of our conscious experience must involve, in some largely unknown manner, the interactions of cortical neurons. However, it seems clear that the ability of brains to form complex patterns depends strongly on the combined actions of EPSPs and IPSPs. Healthy brains apparently require a proper "balance" between these competing processes. The cerebral cortex is also believed to be the structure that generates most of the electric potentials measured on the scalp (EEG). In simple terms, our tissue consists of little bags (cells) of potassium-filled water floating in sodium water, reflecting the distant past when individual living cells floated in the earth's salty oceans. Over millions of years these cells somehow formed new interactions and became larger, multicelled animals. Interactions between neurons now occur as action potentials and the chemical messengers, the hormones and neurotransmitters. Even action potentials constitute drug-delivery systems, but action-potential-generated drugs at the synapses are sharply focused on specific target neurons.

3.7 THE CEREBRAL CORTEX

The cerebral cortex, shown in figure 3-1, consists of the *neocortex*, the outer layer of mammalian brains plus smaller, deeper structures that form part of the *limbic system*. The prefix *neo-* indicates "new" in the evolutionary sense; the neocortex is relatively larger and more important in animals that evolved later. We will use the labels *cerebral cortex*, *cortex*, and *neocortex* interchangeably in this book, unless noted otherwise. The cortex is composed mostly of cells lacking myelin wrapping, and it is often referred to as "gray matter," indicating its color when treated with a special stain by anatomists. But the cortex is actually pink when alive, because of circulating blood.

The cortex contains about 80 percent excitatory neurons and 20 percent inhibitory neurons.[5] The excitatory cells are mostly *pyramidal cells*, neurons that tend to occupy narrow cylindrical volumes like tall skinny trees, as opposed to the more-spherical basket cells (inhibitory) that look more like little bushes. Each pyramidal cell sends an axon to the underlying white matter, a layer of (white) axons just below the cortex. Each pyramidal cell axon connects to other parts of the cortex or to deeper structures. The cerebral cortex surrounds a larger (by volume) inner layer of white matter populated mostly by *myelinated axons*. Such myelin consists of special cells that wrap around axons, making them white and increasing the propagation speed of action potentials, typically by factors of five to ten. Thus, a nonmyelinated axon that propagates signals at one meter per second will typically send signals at five to ten meters per second when myelinated. As discussed in later chapters, this feature of myelin has profound implications for the many brain diseases that attack white matter axons. The cortex exhibits a layered structure labeled I through VI (outside to inside), defined by cell structure; for example, more of the larger pyramidal cell bodies are found in layer V. This layered structure is similar in all mammals.

Figure 3-6 depicts a large pyramidal cell within a *macrocolumn* of cortical tissue, a scale defined by the spatial extent of axon branches (*E*) that remain within the cortex and send excitatory input to nearby neurons. The macrocolumn has a diameter of about three millimeters and a height in the same range. Each pyramidal cell also sends an axon (*G*) into the white matter layer. In humans more than 95 percent of these axons are cortico-cortical fibers targeting the same (*ipsilateral*) cortical hemisphere. The remaining few percent are *thalamo-cortical fibers* connecting to the (sub–white matter) thalamus, or *callosal axons* targeting the opposite (*contralateral*) cortical hemisphere. The probe (*A*) used to record small-scale potentials through the cortical depth is also shown. The dendrites (*C*) provide the surfaces for synaptic input from other neurons, and **J** represents the diffuse current across the cortex resulting from the many membrane current sources and sinks as represented by the expanded picture (*F*). The macrocolumn of figure 3-6 actually contains about a million tightly packed neurons and ten billion or so synapses. If even a small fraction of the neurons were shown, this picture would be solid black. Anatomists are able to identify individual neurons only with a special stain that darkens only a small fraction of the total.

Neuroscience is strongly influenced by the idea that the brain has developed over hundreds of thousands of years. Its evolutionary history is characterized by the successive addition of more parts. In this view, each new structural addition has added more complex behavior, and at the same time imposed regulation on the more-primitive (lower) parts of the brain. The brainstem, limbic system, and cerebral cortex form three distinct levels both in spatial organization and in evolutionary development, as suggested in figure 3-7. In this oversimplified picture, the brainstem, often called the "reptilian brain," or the limbic system might be prevented by the cortex from taking certain actions, say rape and murder, an idea with obvious connections to Freud's id and superego. In more-colorful language, psychiatrists actually treat three patients in each session, an alligator, horse, and a man, representing progressively more advanced brain structures. A more-scientific view retains much of this classical idea but emphasizes that the dynamic interplay between many subsystems in all three evolutionary structures is the very essence of brain function. Such interplay generates dynamic patterns in all of these structures.

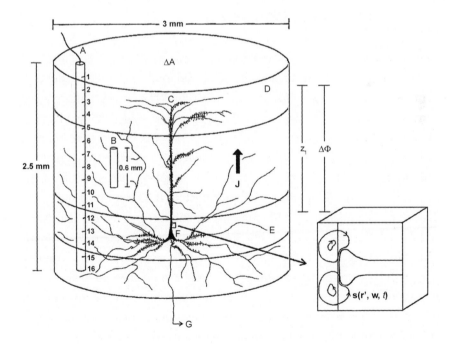

Fig. 3-6. Reprinted from Paul L. Nunez, *Neocortical Dynamics and Human EEG Rhythms* (New York: Oxford University Press, 1995).

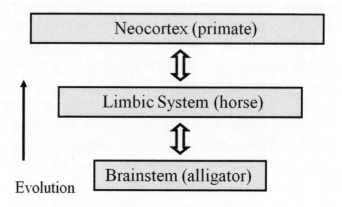

Fig. 3-7.

3.8 THE NESTED HIERARCHY OF THE NEOCORTEX: MULTIPLE SCALES OF BRAIN TISSUE

Here we address one of the most crucial aspects of brain structure and function that it shares with complex social, economic, and other systems— *cortical tissue is arranged in a nested hierarchy.*[6] The hundred billion or so cortical neurons are arranged in columns at various scales defined by anatomical and physiological criteria, as indicated in table 3-1. The *cortical minicolumn* occupies one organizational level up from the single neuron. The minicolumn is defined by the spatial extent of intracortical inhibitory connections; its diameter is approximately 0.03 mm, or roughly the diameter of a human hair. Minicolumns extend through the cortical depth so their heights are about one hundred times their diameters. Each minicolumn contains about one hundred pyramidal cells. Inhibitory connections typically originate with the smaller and more-spherical basket cells. Inhibitory action tends to occur more in the middle (in depth) cortical layers. In some scientific circles, the minicolumn rather than the single neuron is considered the basic entity for neocortical interactions, partly because of the high correlation observed between the electrical activities of different neurons in the same minicolumn, but this issue is context-dependent and open to future studies. Our neurons die naturally all the time; at least several thousand every day. Fortunately, if only a small fraction die in some minicolumn, the minicolumn will probably continue to function normally.

The *cortico-cortical column* (CCC) is defined by the spatial spread of (excitatory) sub-cortical input fibers (mostly cortico-cortical axons) that enter the cortex from the white matter below.[7] CCCs contain about ten thousand neurons and one hundred minicolumns. This extra-cortical input from other, often remote, cortical regions, is excitatory and may spread more in the upper and lower cortical layers, and less so in the middle layers. The selective actions of inhibitory and excitatory processes in different layers may account for changes in local-global behavior in response to different neurotransmitters.[8]

Table 3–1. Multiscale Cortical Columns.

Structure	Diameter (mm)	Number of Neurons	Anatomical Description
minicolumn	0.03	10^2	Spatial extent of inhibitory connections
CCC	0.3	10^4	Input scale for cortico-cortical fibers
macrocolumn	3.0	10^6	Intracortical spread of pyramidal cell
region	50	10^8	Brodmann area
lobe	150	10^9	Areas bordered by major cortical folds
hemisphere	400	10^{10}	Half of the brain

The *cortical macrocolumn* is defined by the local intracortical spread of individual pyramidal cells. As indicated in figure 3-6, pyramidal cells send axons from the cell body that spread out within the local cortex over a typical diameter of about three millimeters. The actual density of axon branches in the cortex is greater than one kilometer per cubic millimeter! If all the axons in a single human brain were laid end to end, they would stretch from New York to Chicago. If even one percent of the neurons in the macrocolumn of figure 3-6 were stained, the picture would be solid black. In addition to the many tree-branch-like intracortical axons, each pyramidal cell sends one axon into the white matter. In humans, most of these reenter the cortex at some distant location. Macrocolumns contain about a million neurons or about ten thousand minicolumns.

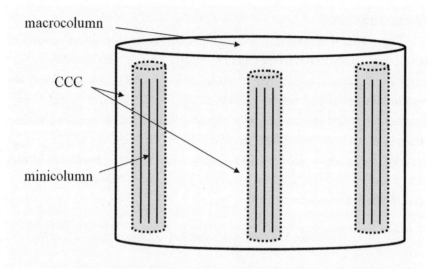

Fig. 3-8.

Inhibitory columnar interactions apparently occur more in the middle cortical layers, and local excitatory interactions are believed to be more common in the upper and lower layers.[9] Such layer-specific physiological properties suggest important implications for cortical dynamic patterns. Healthy brains seem to operate between the extremes of global coherence (all parts of the cortex doing similar things) and functional isolation (each small part doing its own thing). Different neuromodulators tend to operate selectively on the different kinds of neurons at different cortical depths. *Neuromodulators can apparently act to control the large-scale dynamics of the cortex by shifting it between the extremes of global coherence and full functional segregation.* This idea has important implications for several brain diseases, as discussed earlier in this chapter.[10]

An overview of cortical structure reveals neurons within minicolumns, within cortico-cortical columns, within macrocolumns, within Brodmann regions, within lobes, within hemispheres—a nested hierarchy of tissue. The three kinds of nested cortical columns are depicted in figure

3-8. A plausible conjecture is that each of these structures interacts with other structures at the same scale (level) as in neuron-to-neuron interactions or minicolumn-to-minicolumn interactions. But an important point to remember is this: *The rules or procedures of neural interactions may be expected to be different at different scales, as in the case of analogue social systems.* Such differences are anticipated based on the distinct anatomical and physiological features outlined in table 3-1. If inhibitory processes within a column are swamped by excitation (too many EPSPs and not enough IPSPs), this overexcited column may spread its excitation to other columns in the runaway positive feedback process called *epilepsy.*

Cross-scale interactions also occur. Small cell groups produce chemical neuromodulators that act (bottom-up) on neurons in the entire cortex. External sensory stimuli reach small groups of cells in the primary sensory cortex that elicit (bottom-up) global brain responses. For example, this is similar to when your alarm clock goes off and you hop out of bed, begin to plan your day, and so forth. Extended brain networks are activated involving memory and emotion; this global dynamics acts (top-down) to produce action in specific muscle groups. For instance, this causes you to turn off your alarm clock, brush your teeth, put on your pants, and so on. The information storage in memory allowing for all these actions must be enormous; not only for things purposely learned but also for the many other things we take for granted, like how to walk, open doors, and distinguish a mouse from a spouse.

The *morphology* (structure) of the cerebral cortex, in which complex parts occur at multiple scales, reminds us of *fractals*, geometric shapes that exhibit fine structures repeated at smaller and smaller scales. Natural objects that often approximate fractals include coastlines, clouds, and mountain ranges. We need not be concerned here with precise fractal definitions, which include a mathematical property called *self-similarity*, meaning that statistically similar properties occur at multiple levels of magnification. Rather, we note that the cerebral cortex has a fractal-like structure and exhibits fractal-like dynamic patterns, meaning that dynamic patterns exhibit fine structures at multiple scales. Later we will find that measurements of brain activity are very much dependent on the scale of measurement, as expected in nearly all complex systems. For example, intracranial measurements of electrical activity depend on the sizes and locations of recording electrodes.

Top-down and bottom-up hierarchical interactions across spatial scales occur frequently; such phenomena are important hallmarks of complex systems. These terms may sometimes be confusing, especially when they are used differently in different contexts, but recall how they were introduced in chapter 1. We are using the labels "top-down" and "bottom-up" in the context of spatial scales. A small system embedded in a larger system may act locally to produce profound changes in the large system. For example, a small cell group may produce a neuromodulator that acts on many neurons, dramatically changing the global brain state. These are "bottom-up" interactions regardless of the strength or importance of such local inputs. As a result of such bottom-up actions, the large-scale system may be changed in some novel manner like changing global synaptic patterns. The large-scale system may then act "top-down" on some or all of the embedded smaller systems, as in the case of top-down actions of global synaptic action on local cell groups.

Small-scale events cause new events to occur at larger scales, which in turn influence smaller scales, and so forth; in other words circular causality comes into play. In the social analogue system, families, neighborhoods, cities, and nations interact with each other, at the same scales and across scales. National leaders like the president of the United States or chairman of the Federal Reserve exert strong bottom-up influences on humans worldwide. These influences are called "bottom-up" because they are initiated by small-scale entities (single humans); the fact that they are powerful persons substantially increases their influence on the global system, but is still in the "bottom-up" category in the context of spatial scales. At the same time, actions of these leaders are constrained or perhaps even dictated (top-down) by global events like wars and financial crises. One may guess that such cross-scale interactions are essential to brain function, including electrical and chemical dynamic behavior and ultimately consciousness itself.

Similar scientific tools appear appropriate for disparate complex systems; thus, we can reasonably search for more analogous dynamic behavior that may be similar to the brain activity occurring in different states. Cortical analogues provide intriguing hints of possible behavior to be searched for in new experiments—multiscale interactions, standing waves in musical instruments, neural networks, holographic phenomena, chaos, and so forth. The Standard and Poor's 500 stock index provides one brain analogue; the index is composed of markets and associated corporations, made up of separate parts composed of individuals, a multiscale complex system. Same-scale interactions occur between individuals and markets; cross-scale interactions are both top-down and bottom-up. Non-local interactions are commonplace. Variables like prices, profits, index level, and so forth are analogous to multiscale brain measurements. Global economic conditions provide an environment in which the corporations are embedded. The economy acts top-down on corporations, which act bottom-up on the economy. Circular causality is everywhere.

3.9 CORTICO-CORTICAL CONNECTIONS ARE NON-LOCAL

Figure 3-8 depicts cortical columns at the three scales indicated in table 3-1. These columns actually overlap so column boundaries are only determined in relation to designated cells. The axons that enter the white matter and form connections with other cortical areas in the same hemisphere are the cortico-cortical axons, perhaps 97 percent of all white matter axons in humans. The remaining white matter axons are either thalamo-cortical (perhaps 2 percent) connecting the cortex to the deeper thalamus or callosal axons (perhaps 1 percent) connecting one cortical hemisphere to the other. In addition to the long *non-local* cortico-cortical fibers, neurons are connected by *local* intra-cortical fibers that are mostly shorter than a millimeter or so.

With the advent of new imaging technologies, much recent work has focused on constructing the so-called connectome, a comprehensive map of neural connections—essentially the brain's wiring diagram. We will return to this issue in later chapters, especially in the context of white matter disease, but for now, note that brain connections have long been of major interest to at least some neuroscientists. The actual cortico-cortical fibers shown in figure 3-9 were dissected from a fresh human brain immediately after its unfortunate owner was executed for murder in the 1940s.[11] While the donor (reportedly) gave his brain willingly, similar procedures are unlikely to be repeated in today's ethical environment. The average length of the cortico-cortical fibers is a few centimeters (*regional* or *Brodmann scale*), and the longest fibers are roughly 15 to 20 cm (*lobe scale*). The cerebral cortex is divided into about fifty Brodmann areas based on relatively minor differences in cell layers and structures, which in some cases are known to correspond to distinct physiological functions, for example, in the *visual, auditory, somatosensory,* and *motor cortices.* Touch some part of the body, and the matching area of the somatosensory cortex responds; the motor cortex initiates signals sent to muscles. The hundred billion or so cortico-cortical axons are sufficiently numerous to allow every macrocolumn to be connected to every other macrocolumn in an idealized (homogeneous) system. But, because of connection specificity (some columns are more densely interconnected than others), full interconnectivity occurs at a somewhat-larger scale. At the regional scale of several centimeters, we might expect something close to full interconnectivity.

While specificity of fiber tracts prevents full connectivity at the macrocolumn scale, a pair of cortical neurons may be separated by a *path length* of only two or three synapses.[12] That is, an action potential from a neuron targets a secondary neuron, which, in turn, targets a third neuron, and so forth. Only two or three such steps may be required for influences from one region to spread to distant cortices. The cortico-cortical network's path length is analogous to the global human social network with its so-called six degrees of separation between any two humans, meaning a path length of no more than six social contacts worldwide. Pick someone at random, say a Mr. Chen from Taiwan. You probably know someone, who knows someone, and so forth, who knows Mr. Chen, with perhaps only six acquaintance steps needed to complete the path. We humans live in a *small-world* social network. Small-world phenomena are studied in a branch of mathematics called *graph theory* and also appear widely in physical systems. Later we will revisit small worlds in the context of cortical connectivity.

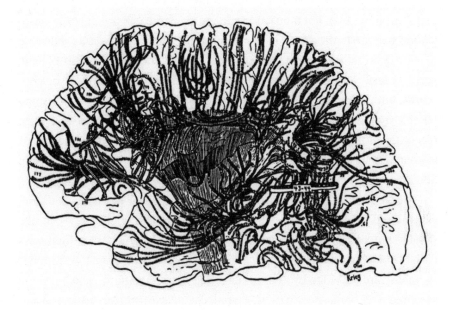

Fig. 3-9. Reprinted from Wendell J. S. Krieg,
Connections of the Cerebral Cortex (Evanston, IL: Brain Books, 1963).

3.10 HUMAN BRAIN EVOLUTION

Our study of consciousness leads naturally to several basic questions about the outer limits of possibilities. Can very simple brains produce consciousness? Is a newborn baby conscious? If so, what about a fetus? Can non-human mammals or even much simpler animals possess consciousness? If so, how can we know? My golden retriever, Savannah, appears to possess at least rudimentary consciousness. Two-way communication between us is commonplace; she obeys commands and lets me know when she wants to chase a ball by "impatiently" nudging my hand, holding the ball. At various times she exhibits external behavior that we humans associate with excitement, sadness, fear, and even guilt. The dog provides an especially interesting case of apparent nonhuman consciousness, partly because of the domestic dog's evolutionary history with humans. Natural selection has favored dogs that communicated efficiently with their human companions over more than ten thousand years. A plausible guess is that this association has made the dog's internal experience more humanlike than that of a wolf or perhaps even a chimpanzee in the wild. It's sometimes said, partly in jest, that philosophers who own dogs believe their dogs to be conscious, but those philosophers who don't own dogs think not. Establishing reliable quantitative measures of consciousness is no easy task. However, recent fMRI and other studies of dog brains have led some neuroscientists to conclude that dog consciousness occurs at a level similar to very young human children. Maybe Savannah is really just a zombie, but I doubt it.

We know that consciousness is far from an all-or-nothing condition, as explored in more detail in chapter 4. Consciousness does not work like a light switch that just goes on and off. Rather it's more like a light with variable brightness controlled by a dimmer switch. In brains, the "dimmer switches" are provided by neurotransmitter, neuromodulator, and hormone systems. We humans experience graded states of consciousness when drowsy, under the influence of alcohol, dreaming, and in various stages of mental illness or Alzheimer's disease. Furthermore, our consciousness and free will are strongly influenced by a number of partly independent preconscious systems. In order to investigate the multiple levels of human and nonhuman consciousness, some provisional quantitative framework will prove helpful. To this end we turn to a layered model of consciousness proposed by neuroscientist Antonio Damasio, a view based on three progressive developmental stages.[13] For now, we tentatively adopt his framework supporting various categories of consciousness, but our future discussions will not be confined by this or any other model. Damasio's three hierarchical layers are deeply anchored in evolutionary considerations:

- *The protoself* represents a pre-conscious state shared by all life-forms including single-celled organisms. These entities detect internal and external physical changes that affect the organism's well-being. The protoself provides an essential foundation for the two higher stages of "self" to build on.
- *Core consciousness* occurs when organisms develop a sense of self

as brains build representative images; this level is not exclusive to humans. Core consciousness is concerned only with the present moment, the here and now. It does not require language or memory; nor can it reflect on past experiences or project itself into the future.

- *Extended consciousness* occurs when awareness moves beyond the here and now and employs extensive use of memory. This so-called autobiographical layer of self is anchored in the more-basic protoself and core consciousness. It gradually becomes aware of past, present, and future. While the extended consciousness may be substantially enhanced by language, language is not necessarily required.

With these categories in mind, one can guess that mammals, birds, octopuses, and many other animals possess core consciousness, but perhaps most animals lack the extended consciousness of humans. The question of whether higher animals like the nonhuman great apes (orangutans, gorillas, and chimpanzees) possess some lesser (but nevertheless extended) consciousness will likely remain open to question for the foreseeable future. Genetic, anatomical, and behavioral aspects link us to other animals, especially the other great apes; for example, chimps share about 95 percent of their DNA with humans. Chimps, and even crows and octopuses, use tools, solve problems, and exhibit many other human-like behaviors.

In the context of brain anatomy, we may ask, what distinguishes the human brain from the brains of other mammals? This question presents an apparent paradox—cognitive processing, indeed most of our conscious experience, depends critically on the dynamic operations of the cerebral cortex. But there is very little difference between the cortical structures (*morphology*) of different mammals; all contain similar cell types and distinct cortical layers. This striking similarity even includes a constant number of cells within the minicolumn. Larger brains will, of course, have more neurons and columns to interact with one another. While brain size surely counts for something, it fails to explain why humans seem more intelligent than elephants, dolphins, and whales, at least based on our biased human measures. Whales have not developed advanced technologies or even constructed moderately sophisticated tools, but maybe they just have better things to do. These three large species all have brains covered by folded cortices that look much like human brains, except they are larger—*much* larger in several whale species. While humans do enjoy the largest ratios of brain weight to body weight, such a measure contributes little to the fundamental question of why human brains produce much more complex behavior. I have gained weight since high school, but I don't think this has made me dumber; rather it's probably the other way around.

The compelling issue of brain humanness may be addressed from several different directions. How does human brain anatomy differ from that of other mammals or lower animals? How does human behavior differ? To address these questions, first consider just where brains came from in the first place; that is, a brief look into evolutionary history is suggested. The human brain is the product of a vast evolutionary development. The earliest life on Earth arose at least 3.5 billion years ago. From fossil records, scientists infer that the first brain structures appeared in worms over 500 million years in the past. Dinosaurs and early mammals appeared at around minus 200 million years. Anatomically modern *Homo sapiens* seem to have first appeared much later, probably in Africa around minus 200,000 years. By minus 50,000 years, humans had migrated to most other parts of the world. The first civilizations emerged about 5,000 years ago in locations that are now parts of India, China, Egypt, and Iraq.

The following list puts these vast time intervals in human perspective by representing the entire period of life on Earth as a single year, thereby compressing all of human history and prehistory into just 30 minutes:

- January 1. First known life appears in the form of single cells.
- March 7. First worm brains appear.
- December 10. Dinosaurs and early mammals appear.
- December 18. Last common ancestor of humans and chimps disappears.
- December 24. Massive extinction event kills three-quarters of Earth's plant and animal species, including the non-avian dinosaurs. This extinction is believed due to a massive asteroid that hit what is currently known as the Gulf of Mexico, possibly creating

an extended global winter that blocked the sun's energy from plants and plankton.

- December 31, 11:30 p.m. First anatomically modern humans appear in Africa.
- December 31, 45 seconds before midnight. First human civilizations appear in the Middle East.
- December 31, one-half second before midnight. First human walks on the moon.

The story of human evolution reveals many subplots. One subplot involved with brain development concerns the Neanderthals, a subspecies of *Homo sapiens* that inhabited Europe for a long time before the arrival of modern humans. The two groups apparently coexisted for more than five thousand years and crossbred. As a result, Neanderthals contributed to the DNA of modern humans, including most non-Africans as well as a few African populations, probably about fifty thousand years ago.

Until recently, Neanderthals were often portrayed as dumb brutes in popular culture. Neanderthals had a larger-than-average brain and body size compared to modern humans. Archaeological evidence indicates that they probably communicated by speech and used tools. More-recent artist renderings of Neanderthals have become much more intelligent-looking, closely resembling modern humans. But why did the Neanderthals die out while modern humans survived, all the while living near humans for several thousand years? Several possible answers have been advanced. Perhaps the modern humans (*Homo sapiens sapiens*) arriving later in Europe from Africa carried some fatal disease that Neanderthal immune systems could not handle. Or perhaps modern human brains had developed much more sophisticated language functions, allowing for more effective cooperation between individuals and the formation of more highly functioning tribes. In later chapters, we will revisit the essential language issue in the context of human consciousness. Still other influences have been suggested by anthropologists, but we may never know which ones were the most pivotal to human natural selection.

3.11 WHAT MAKES THE HUMAN BRAIN "HUMAN"?

The astounding human evolutionary success, in which human activity currently dominates the earth's ecological systems, is typically attributed to a relatively large brain with a more-extensively developed neocortex, especially prefrontal cortex and temporal lobes. These brain structures are believed to play critical roles in higher levels of reasoning, problem solving, language development, and interactions with other humans. According to the standard story, these anatomical developments allow humans to use and develop tools more often than any other animal. Humans even construct sophisticated machines; they also develop mathematics, technology, art, and much more. Yet the relatively minimal anatomical changes that separate chimps from humans seem quite modest in comparison to the enormous technological and behavioral advances of the human race. A plausible conjecture is that these relatively small anatomical changes constitute only a small part of the human brain's evolutionary story.

Figure 3-10 is a famous drawing from one of Charles Darwin's books,[14] which includes these words in the caption: "chimpanzee disappointed and sulky," implying a quite-high level of consciousness in chimps. Communication between humans and chimps is moderately successful but far more limited than intra-human exchanges are. Robust language communication is an especially important aspect of human social interactions and tribe formation. The development of human language has been a hot scientific topic for many years, but no consensus on either the origin or age of human language is evident. The emergence of language lies far back in prehistory; the relevant developments have left no direct historical traces. Humans have established many systems of symbolic communication for self-expression and the exchange of ideas. And they create complex social structures composed of cooperating and competing groups often arranged in nested hierarchies. With the critical bottom-up assistance from language skills, human social structures have become progressively more complex, leading only relatively recently to bona fide civilizations. We may guess that the emergence of tribes and other social structures provided important top-down influences to accelerate language development in individuals.

Such circular causality in language development may be a central factor in making the human brain "human."

Fig. 3-10.

To follow up on the problem of "brain humanness" in an anatomical context, note the modern idea that the dynamic interaction between brain subsystems lies at the very essence of brain function. This view suggests an emphasis on dynamic patterns driven by interconnections between subsystems rather than just the anatomy of the subsystems themselves. Some neuroscientists have emphasized an important quantitative difference in the brains of different species of mammals. Suppose we count the number of axons entering and leaving a patch of the underside of the neocortex. Some fibers connect cortex to cortex (cortico-cortical fibers); others connect cortex to deep (midbrain) structures, especially the thalamus. Some years ago, one of my colleagues consulted the anatomical literature to estimate the following ratio for a typical patch of neocortex:[15]

$$Ratio = \frac{number\ of\ cortico\text{-}cortical\ fibers}{number\ of\ thalamo\text{-}cortical\ fibers}$$

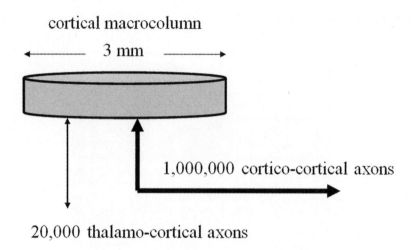

cortical macrocolumn

<-------- 3 mm -------->

1,000,000 cortico-cortical axons

20,000 thalamo-cortical axons

Fig. 3-11.

In the human brain, this ratio is large, perhaps in the range of roughly 20 to 50, meaning that only about 2 percent to 5 percent of human fibers entering or leaving the cortex connect to midbrain structures. As indicated in figure 3-11, a cortical macrocolumn of ten square millimeters sends about one million axons to other parts of the cortex, and it receives cortical input from about the same number. By contrast, only about twenty thousand fibers connect to the thalamus and other midbrain structures. In contrast to this human anatomy, the relative density of thalamo-cortical compared to cortico-cortical fibers is apparently substantially higher in other mammals. Data from several mammals are summarized in figure 3-12; the vertical axis provides estimates of the ratio defined above. This histogram suggests that the fraction of cortico-cortical fibers becomes progressively larger as mammals become capable of more complex behavior. Unfortunately, we do not have corresponding estimates of this ratio for elephants, dolphins, and whales, data which may or may not support these ideas.

The estimates of figure 3-12 provide an intriguing, but tentative and incomplete, explanation for human brain *humanness* that makes intuitive sense in several ways. I claim that I am smarter than my dog, Savannah, but in what sense am I smarter? My olfactory cortex is a moron compared to Savannah's. She experiences a whole world of distinct smells that I can barely imagine; I go sightseeing, she goes smell-sniffing. She finds golf balls in tall grass and under water by *smellor* technology, the doggie version of infrared sensing. In some ways that we don't understand, our humanness seems to originate largely from global interactions of multiple cortical neurons and columns at different scales within the nested hierarchy of cortical tissue. Dynamic feedback linking cortex to cortex may be relatively more important in humans than in lower mammals, but this does not negate the critical influence of midbrain chemical (neuromodulator) control acting over longer timescales. We seem to be more humanlike than horselike or alligatorlike—at least most of us do.

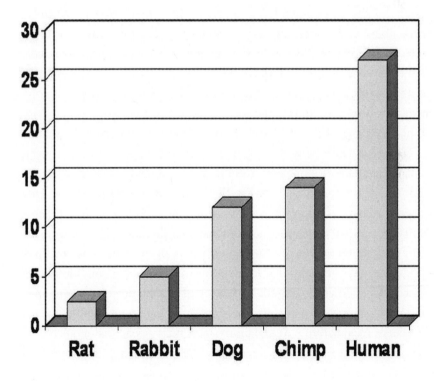

Fig. 3-12. A modified version of the original figure produced by Ron D. Katznelson, "Normal Modes of the Brain: Neuroanatomical Basis and a Physiological Theoretical Model," in Paul L. Nunez, *Electric Fields of the Brain: The Neurophysics of EEG*, 1st ed. (New York: Oxford University Press, 1981), pp. 401–442.

3.12 SUMMARY

I have emphasized the special features of brain anatomy and physiology that facilitate complex dynamic patterns in this chapter. Features typically common to complex systems include nested structural hierarchies, scale-dependent observations, non-local interactions, circular causality, emergent phenomena, and unpredictable responses to external environments. One useful brain metaphor is the "Christmas-tree brain." Electrical, metabolic, or other measures "light up" in certain brain regions. While this picture embodies several useful ideas, the metaphorical Christmas tree provides only a very impoverished brain model.

The central nervous system consists of brain and spinal cord. The peripheral nervous system has two parts—the somatic system consists of nerve fibers in the skin, joints, and muscles under voluntary control, our apparent expressions of free will. The autonomic system includes nerve fibers that send and receive messages from internal organs, thereby regulating unconscious processes. Much of our conscious experience involves the interaction of neurons in the cerebral cortex, which also generates most of the electric (EEG) and magnetic (MEG) fields recorded on the scalp.

The distinction between dynamic states of functional isolation versus global coherence is emphasized. Subsystems sometimes interact strongly such that the population behaves like a single integrated entity. Other systems consist of many little subsystems that act more or less independently. A football fan analogue demonstrates this idea. Before the game begins, most interactions occur between fans sitting close together in the stadium, a state of functional localization. The opposite condition occurs when a touchdown is scored, and most of the home team's fans cheer together, a state of global coherence, but still containing small embedded pockets of local activity. Actual brain states that appear analogous to these metaphorical states include schizophrenia and autism at the localized end, healthy states of moderate localization, and coma or anesthesia states of extreme global coherence.

Sensory input to a brain may or may not elicit an output, typically a perceptual or physical response. We know quite a bit about the input and output ends of this gray box, but much less about the workings of regions in between. Human brains are made up of hundreds of substructures, including the gray matter of the cerebral cortex, and the deeper white matter layer, consisting mostly of cortico-cortical fibers connecting different parts of the cortex to itself. The high density of these long axons in humans compared to lower mammals may be an important aspect of our humanness.

Neurotransmitters are special chemicals stored in the synaptic vesicles located at the pre-synaptic side of synapses. When electrical signals arrive at the end of axons, transmitters are released into the gap between neurons where they bind to specific receptors in the target neuron. These chemical messengers play a major role in shaping our everyday lives. Faulty neurotransmitter systems cause a number of brain diseases. Drugs work by interacting chemically with specific neurotransmitters and may shift brain states along the local-to-global spectrum.

Neurons and cell assemblies communicate both electrically and chemically. The neuromodulators are a class of neurotransmitters regulating widely dispersed neural populations. Like hormones, neuromodulators are chemical messengers, but they act only on the central nervous system. By contrast to targeted synaptic transmission of neurotransmitters, neuromodulators diffuse through large areas of the nervous system with global effects on vast numbers of neurons.

While chemical control mechanisms are relatively slow and long-lasting, electrical events turn on and off much more quickly. Electrical transmission over long distances occurs by means of action potentials that travel along axons in the peripheral nervous system, cortico-cortical axons, and on nonmyelinated axons within cortical tissue. Inhibitory and excitatory processes tend to occur in selective cortical layers. Similarly, different neuromodulators tend to operate selectively on the different kinds of neurons at different depths. Healthy brains seem to operate between the extremes of global coherence and functional isolation. Neuromodulators control the dynamics of the cortex by providing for shifts between global coherence and functional isolation.

Experience with other complex systems suggests that the nested hierarchical structure of cortical columns may have profound influence on cross-scale interactions, both top-down and bottom-up. Inhibitory processes may allow functional segregation of selected cortical tissue, effectively walling off portions of tissue that participate in discrete networks. Neural networks are expected to form at multiple spatial scales and operate over multiple timescales. Non-local interactions by means of the cortico-cortical fibers apparently allow for more-complex dynamic patterns than would be possible with only local interactions. In the following chapters, we will explore various means by which such structural properties may influence or perhaps create our different mental states.

Chapter 4
STATES OF MIND

4.1 WHAT IS CONSCIOUSNESS?

C an we even define consciousness in a coherent manner? Yes, sort of, but our definitions are bound to be circular: *Consciousness is the state of awareness of an external environment or of something within oneself; the ability to experience or feel.* OK, but what is "awareness" other than a state of consciousness? Apparently, we are stuck with circular definitions. Can we ever hope to understand consciousness? We must be careful with the often misused word *understand.* When returning graded exam papers, we science professors are often approached by students uttering a classic line, "I understood everything you said in class, but I could not solve the exam problems." My personal response to such students is to assume my best "guru posture" and reply, "There are many levels of understanding, my son; your job is to reach for deeper levels." To cite another example, suppose I ask if you understand how your television works? "Sure," you answer, "I press this button and it goes on or I push that button to change the channel." The point of these little stories is this—we should avoid the label "understand" when all we really mean is that we are comfortable with something that is part of our daily routines. While conscious experience is the most familiar aspect of our lives, it is also the most mysterious. A skeptical (and humorous) definition of consciousness was provided by the British psychologist Stuart Sutherland in the 1989 version of the *Macmillan Dictionary of Psychology*:

Consciousness—the having of perceptions, thoughts, and feelings; awareness. The term is impossible to define except in terms that are unintelligible without a grasp of what consciousness means. Many fall into the trap of equating consciousness with self-consciousness—to be conscious it is only necessary to be aware of the external world. Consciousness is a fascinating but elusive phenomenon: it is impossible to specify what it is, what it does, or why it has evolved. Nothing worth reading has been written on it.[1]

While your author does not endorse the last sentence in this entry, it does a nice job serving as an "in your face" summary of the issues as many intelligent people see them. Philosophers since the time of René Descartes and John Locke have struggled with the consciousness challenge by attempting to pin down its essential properties. Can consciousness ever be explained in materialist or reductionist terms? Or does it demand a dualistic distinction between mental and physical states? Does consciousness exist in lower animals? Are conscious computers or robots theoretically possible? If robot consciousness were to be claimed, how could this claim be verified or falsified? Can we come up with good quantitative measures of consciousness? That is, can we ever say that some mental state is X percent more conscious than another state, or that some higher animal is Y percent more conscious than some lower animal?

A plausible, but preliminary, attempt at consciousness quantification was outlined in chapter 3 with Antonio Damasio's three hierarchical levels of consciousness—the *protoself, core consciousness*, and *extended consciousness*. "Here and now" awareness of the environment occurs in core consciousness. Self-awareness and knowledge of relationships between past, present, and future require extended consciousness. An important issue left open thus far is whether the very highest levels of consciousness (we might label them "hyper-extended consciousness") require language, and maybe substantial social interactions. Later we will return to these fascinating issues of language, social contacts, and possible extended consciousness.

Consciousness is now considered to be a bona fide research topic in psychology and neuroscience. The primary focus is determining the neural and psychological correlates of consciousness, the so-called easy problem. The hard problem of the origin of consciousness itself is often dismissed but may be tentatively approached through studies of the easy problem. In other words, finding consciousness correlates will not, of itself, solve the hard problem; however, such "easy" information is expected to constrain and mold our ideas about the hard problem. Simply put, consciousness requires the brain, not the liver. People are conscious, but rocks or trees are not. In this chapter we continue our discussions of relationships between the brain's dynamic patterns and various states of mind.

Most experimental studies assess consciousness by asking human subjects for verbal reports of their experiences. Other approaches employ masking techniques to study unconscious and pre-conscious processes. In hospitals and other clinical settings, consciousness is assessed by observing a patient's arousal and responsiveness. Consciousness presents itself as a continuous range of medical states from full alertness and comprehension, through disorientation, delirium, loss of meaningful communication, and even the absence of movement in response to painful stimuli. Immediate issues of practical concern include assessment of the consciousness in severely ill, comatose, or anesthetized patients and how to treat impaired consciousness.

4.2 WHAT IS IT LIKE TO BE A BAT?

The title of this section refers to a widely cited and influential thought experiment proposed by philosopher Thomas Nagel.[2] He chose the bat as the star of his essay because bats seem to operate more or less near the midrange of the consciousness spectrum. Bats are mammals; so most of us probably think they possess at least some rudimentary elements of core consciousness. On the other hand, their life experiences differ substantially from those of humans or even higher mammals like the great apes. You and I can perhaps imagine what it would be like to fly, navigate by sound waves, hang upside down, and even eat bugs. But our imaginations are still severely limited because we cannot escape the limits of our subjective perspective. We can never really experience "batism" directly. Nagel argued that consciousness cannot be explained without reference to the subjective character of experience. Furthermore, this subjective nature cannot be explained by reductionism. Thus, Nagel advanced an argument against most of the categories of materialism and reductionism outlined in chapter 2.

Nagel's arguments emphasize the importance of the *explanatory gap*, the difficulty that materialism has in explaining how physical properties can give rise to conscious minds. We suggested earlier that materialism appears to be inconsistent with the metaphysical existence of philosophical zombies, those imagined creatures that behave just like conscious humans but actually lack consciousness. The basic argument says essentially that such zombies are missing some essential aspect of humanness; we call it the "C-factor" leading to consciousness. The existence of genuine free will also argues in favor of C-factor existence. According to our basic zombie definition, no human tests can ever measure or record the C-factor. Thus, it seems to operate within a "layer of reality" that is forever hidden from humans. In other words, some "shadow world" or strange informational structure seems to be implied that influences our observed universe in ways that perhaps we can never hope to understand or even submit to scientific scrutiny. Later we will introduce the concept of *ultra-information* in this context.

"Wait just a second!" you may say. "I thought I was reading a book anchored in genuine science; not some flaky mumbo-jumbo New Age drivel associated with astrology, quantum mysticism, or whatever." But in truth, the notion that the so-called empty space of a vacuum has fundamental properties is mainstream science. Tiny virtual particles and electric fields jump in and out of existence; space is warped by massive bodies; mass and information transfer through space are limited to the speed of light; and so forth. Thus, there should not be much argument about the existence of some sort of shadow world—in truth, an extensive controversy about its actual nature is robust and ongoing. One thing seems clear: the properties of the particular space-time in which we live have allowed consciousness to emerge. Most philosophers and scientists probably agree that alternate and/or parallel universes containing life but no consciousness are metaphysically possible. To this scientist and wannabe amateur philosopher, such a line of thinking implies that our particular space-time embodies some sort of hidden information or structure or something else that is absent in the imagined "unconscious universes."

Admittedly, we are treading on murky scientific ground with these philosophical speculations. Taken to extremes, such reasoning may lead to a slippery slope, possibly dumping us into the stormy sea of extreme *panpsychism*, the position that the fundamental physical components of our universe have mental properties. This philosophical position is quite different from the much milder claim that our universe has fundamental properties that have allowed consciousness to develop, and that such properties could well be absent in alternate universes, perhaps even the life-friendly ones. At least in this limited context, some sort of proto-consciousness seems to be a fundamental property of our universe, perhaps in roughly the same sense that electric charge is fundamental. Such a modest claim does not insist that rocks or thermostats or even powerful computers possess rudimentary consciousness.

So, my scandalous message is this—a shadow world of some kind has been confirmed by mainstream physics. But, this shadow world may or may not have much of anything to do with consciousness. We will revisit this subject matter in later chapters, but for now suffice to say that quantum mechanics, the pre-eminent field of modern physics, has been confirmed by innumerable experiments and practical applications in electronic technology. Quantum ontology relies on the existence of a shadow reality of some sort, but nobody is sure about its actual nature. In various interpretations, the proposed shadow reality is associated with labels like *many worlds*, *implicate order*, *wave-function collapse*, *uncertainty principle*, *superposition of states*, *Platonic world*, *spooky action at a distance*, and so forth.

Yes, I have pledged to limit our speculations to those that do not clearly violate established physical laws. But, it turns out that this pledge is not as constraining as some readers might think. Some weird stuff will soon appear in our headlights, and many classical ideas will be exposed as passé or even dead wrong. We will advance the argument that the happy marriage of complexity science to neuroscience may be in danger of being constrained by outdated physical ideas. The revolutionary quantum world revealed in the early 1900s shocked physicists of the time, but today these ideas constitute mainstream physics. As we have emphasized, becoming familiar with common entities or occurrences is not the same thing as deep understanding, not even close. Physics, it turns out, has its own versions of the easy and hard problems. If we are really serious about answering the consciousness challenge, we better get used to these strange ideas. Close scrutiny of the physical world that we inhabit will prove essential, even if we later end up sticking with strong reductionist or materialist positions on the consciousness question.

4.3 PATTERNS AND SUB-PATTERNS OF CONSCIOUSNESS

Brain science has shown unequivocally that the space-time behavior of various kinds of dynamic patterns provides useful measures of brain function; these patterns are linked to human behavior and consciousness. These dynamic patterns are neural correlates of consciousness—patterns of electric fields, blood-oxygen levels, metabolic activity, and more. EEG patterns indicate sleep stages or depths of anesthesia; fMRI reveals face recognition, and so forth. We have defined the label *dynamic pattern* quite broadly here to encompass brain measurements that are currently in play, as well as different kinds of patterns that may be measured in the future. The adjective *dynamic* emphasizes that pattern time dependence is generally just as important as its spatial dependence. Sound waves, quantum waves, electromagnetic fields, stock market prices, global air temperatures, protein molecules, and neurotransmitters all occur as dynamic patterns even though we can't see many of them.

The word *pattern* implies something that is spread out spatially. Single points like the dot at the end of this sentence don't normally exhibit patterns. However, entities that at first appear to be simple points can turn out to consist of complicated patterns when magnified. These kinds of patterns often fall into a category called *fractals*, objects that exhibit detailed physical structure at many levels of magnification. Figure 4-1 depicts a large-scale network metaphor; the arrows labeled "global field" represent the top-down influence of the environment in which the networks are embedded.

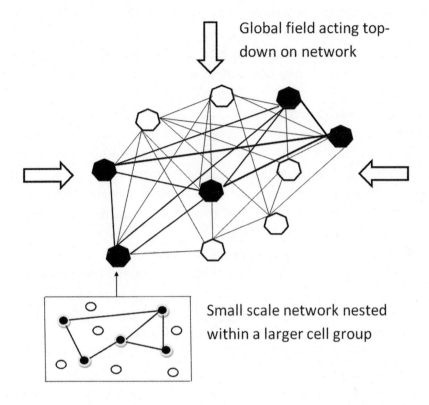

Global field acting top-down on network

Small scale network nested within a larger cell group

Fig. 4-1.

The network nodes are represented by small seven-sided figures (hep-tagons) to signify some sort of internal structure. These nodes might rep-resent tissue presented at any number of spatial scales; for example, we might say these are macrocolumns or cortical regions. Each macrocolumn contains about one million neurons and ten thousand minicolumns; each cortical region contains about a hundred macrocolumns. Thus, we can envi-sion networks within networks, and so forth; we might reasonably expect multiscale dynamic patterns to come along for the ride. Analogue and met-aphorical networks, rather than genuine neural networks, are employed in our discussions. This choice remains faithful to our goal of dealing with basic ideas without becoming bogged down in technical details. However, our analogue networks mirror a number of genuine brain networks that have been tracked at different spatial scales as nicely presented in the work of Olaf Sporns, *Networks of the Brain*.[3]

Nature's dynamic patterns are seen just about everywhere, reminding us that brain pattern measurements of any kind are limited to their own characteristic ranges of resolution. Stick an electrode of a certain size into the cortex—the recorded electric field represents a space average over the volume of the electrode tip; spatial resolution is strictly limited in this manner. Or, employ fMRI to measure blood oxygen in brain tissue; the resulting images represent time averages over minutes or seconds, as well as space averages over little tissue volumes (voxels). Spatial and temporal resolutions are always limited in the real world of genuine experimental science. The critical modifier *dynamic* in "dynamic pattern" indicates that consciousness signatures change over time, but each conscious "snapshot" actually represents a process spread out over about one-half second. Our experience of the "now" really refers to an average over several hundred milliseconds in the past. Consciousness has been colorfully and accurately tagged as the *remembered present* by Gerald Edelman.

Chapter 3 emphasized that the structure of the cerebral cortex consists of complex parts that occur at multiple spatial scales—the cortical columns and so forth. This fractal-like structure suggests that the various kinds of spatial patterns measured by neuroscientists should also be strongly scale-dependent. Similarly, top-down and bottom-up hierarchical interactions across spatial scales are important hallmarks of most-known complex systems. Small-scale events cause new events to occur at larger scales, which in turn influence smaller scales, and so forth; that is, circular causality is found almost everywhere in the natural world.

A plausible conjecture is that such cross-scale interactions are essential to the appearance of consciousness. This tentative idea is based first of all on the anatomical fact of the brain's nested structural hierarchy—we might naturally ask why evolutionary forces have organized tissue in this way unless such nested structure contributes significantly to brain function, consciousness, or both. Second, conscious brains and many better-known complex systems seem to require some minimal level of complexity based on multiscale interactions. This implies that the kinds of brain dynamic patterns that scientists have measured thus far may constitute only a very tiny fraction of the big picture. From the lofty position of a god's-eye view we might expect consciousness correlates to consist of patterns within patterns within patterns and so on over a wide range of spatial scales. Somewhat-similar arguments apply to multiple timescales, but we will hold off discussion of time resolution until dealing with the important issue of *brain rhythms* and their close connection to mental states in chapter 6.

4.4 THE CONSCIOUS AND THE UNCONSCIOUS

Various categories of consciousness are summarized in figure 4-2, where two-way communication with normal humans occurs only for brains to the right of the dashed center line.[4] This picture ranks brain states approximately in terms of their ability to produce physical action (*motor functions*) and mental functions. The labels mostly represent recognized clinical conditions but include several gray areas and controversies; nevertheless, this picture provides a reasonable introduction to multiple conscious states.

- The *healthy alert* person (upper right corner) operates at high awareness and physical functions, fully able to plan and carry out action. When *drowsy*, his muscle and mental skills are impaired. By contrast, under *hypnosis* he can concentrate intensely on a specific thought, memory, or feeling while blocking distractions. The *right brain* label refers to the apparent separate consciousness of an epilepsy patient having undergone the split-brain operation. We will later return to this topic.
- Most dreaming occurs in *REM sleep*, so labeled because of the rapid eye movements typically visible under the subject's closed eyelids. Dreaming is a kind of limited consciousness, one with minimal connections to waking reality. What good are dreams? No one knows.
- In *coma*, *anesthesia*, and *deep sleep* we are mostly unaware and lack voluntary movement. The caveat "mostly" is based on evidence that the unconscious mind can store some information during these states.

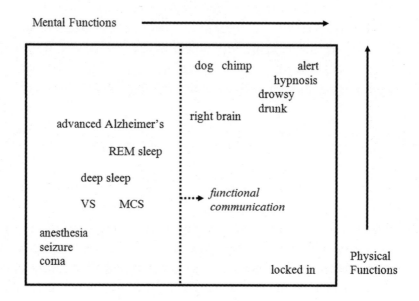

Fig. 4-2.

- *VS* is the clinical category called the *vegetative state* associated with severe brain injury, often caused by accident or stroke. Some patients "awake" from coma, open their eyes, but remain unconscious. Wakefulness in this case means only that vegetative patients exhibit sleep and waking cycles revealed by EEG.
- Patients who recover from the vegetative state may gradually start making deliberate movements but remain unable to express thoughts and feelings. This level of functional recovery is categorized as MCS, the *minimally conscious state*. Like the vegetative state, the minimally consciousness state may be a transient condition on the way to full or partial recovery, or it may be permanent.
- *Alzheimer's disease* covers a broad spectrum of consciousness, ranging from near normal in early years to unconsciousness, perhaps years later. This cruel disease causes those afflicted to gradually "lose themselves" over long times.

- The *locked-in syndrome* is perhaps the most disturbing of all medical conditions. The locked-in patient may be fully conscious but cannot move or communicate due to paralysis of all or nearly all voluntary muscles. The majority of locked-in patients never regain voluntary movement.

4.5 SLEEP

We are all very familiar with the loss of consciousness that occurs during sleep, especially the deep sleep that tends to occur early in the night. Sleep stages and waking conditions are controlled by *neuromodulators*, the special chemicals released by cells deep in the brain; the largest contributions probably come from cells in the brainstem. The neuromodulators influence the excitability of target cells in other parts of the brain. As a result of this control by lower brain structures, the brain and mind cycle through various stages of sleep every night. The distinct stages are clearly revealed by recordings of eye movements, muscle tension, and the electroencephalogram (EEG), as discussed in more depth in chapter 6. The sleeping brain is capable of generating an imaginary inner experience, a dream world that is disconnected from the real world. Even during sleep, parts of our pre-conscious selves remain somewhat alert. For example, the sound of your baby crying or hearing your name called are often effective waking signals.

In the repeating stages of human sleep, the body alternates between *non-REM* and *REM sleep*. REM stands for "rapid eye movement," but this stage also involves other parts of the body, including near paralysis of muscles, thereby preventing sleepwalking in most people. REM sleep almost always occurs with dreams. Non-REM sleep is itself divided into stages 1, 2, and 3, indicating depth of sleep. Stage 1 indicates the drowsy period between sleeping and waking; some people also experience hallucinations during stage 1 sleep. Stage 2 is the intermediate depth of sleep; stage 3 is the deepest stage. Figure 4-3 represents sleep stage pathways during a night's sleep with arrows indicating typical transition directions.

At one time or another, perhaps after a night of partying, most of us have been rudely awakened by our alarm during stage 3 sleep. If this happens, you will probably know it because of the strong feeling of not wanting to get up for any reason. The house is on fire? So what; you just want to sleep. You may suffer *sleep inertia*, a mental deficit that can last for ten minutes or so after waking from stage 3 sleep. While dreams nearly always occur during REM sleep, you won't remember most of them the next morning. However, if you are awakened during REM sleep, you will probably remember dreams in some detail. Dreams are not confined solely to REM sleep; dreamlike experiences can occur during any stage of sleep, but they typically lack the rich content of REM sleep.

Dreams present a strange kind of consciousness in which we remain disconnected from the outside world. Normally, the dreamer has no control of his private experience; he is a passive spectator like someone watching a movie. But sometimes the dreamer can become aware that he is dreaming; this is called a *lucid dream*. In such cases the dreamer may be able to remember aspects of his waking life and may even be able to direct his dream. Your author has tried to direct dreams several times but has failed miserably. One night I was experiencing a very pleasant dream, but suddenly "awoke" to find myself walking through a shopping mall. I realized right away that this was impossible so I must still be dreaming. My first dream state was a dream within a dream. I thought to myself, "Hey, this could be fun; now I can do anything I want with no consequences to myself or anyone else. I know; I'll fly around the mall." Unfortunately, I could not manage to levitate even a single foot; I was just as grounded in my dream as I am in real life. Next, I noticed two women carrying trays of food in the mall. I went up to them and knocked their food trays on the floor. Somehow I knew that the pair consisted of a mother and her grown daughter. I wondered why I did such a cruel thing to this nice couple; I even felt a bit guilty. End of dream.

Sleep has been studied for a long time with EEG and with several kinds of direct observation of subject behavior. Erections of the penis and clitoris typically occur in normal REM sleep, even with non-erotic dreams, along with greatly reduced muscle tone over the entire body. Reduced muscle tone usually prevents sleep walking and other activities. We can dream away in this stage, but not much physical action is likely to come of it. Each of us has perhaps four or five periods of REM sleep per night, one or two hours in total, but most dreams are not remembered in the morning. Dreams are rich in emotion, especially fear and anxiety. Most dreams deal with personal concerns, such as being inappropriately dressed or not dressed at all, being lost, or being late for a college exam. Exam dreams may occur years after graduation; dreams are often stuck in the past. Pre-sleep experiences like viewing horror movies just before bed rarely influence dreams. Again, what good are dreams? Nobody is really sure.

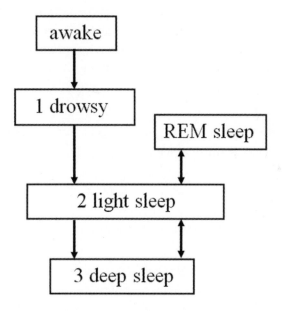

Fig. 4-3.

Sleep disorders are commonplace. Persons deprived of sleep can often undergo unexpected short periods (maybe five to ten seconds) of stage 1 sleep. Such episodes can be dangerous when constant attention is needed, as in driving a car. I recall a scientific meeting on brain-computer communication where I met a young man permanently confined to a wheelchair because of such a tragic sleeping event. In other cases, sleepwalkers walk around with eyes open and sometimes speak slowly. Occasionally they become agitated—screaming, running, and displaying aggressive behavior. In a few cases, sleepwalkers have actually committed "murder," creating tricky legal questions. Sleepwalking normally occurs during stage 3 sleep. Sleep talking occurs more often than sleepwalking and can occur in both non-REM and REM sleep. *REM sleep disorder* afflicts mostly elderly men; it involves violent episodes of dream enactment, often with punching, kicking, and leaping from bed. These events occur during REM sleep, and the dreamer can usually recall the dream in detail. A man may dream that he is defending himself or his wife, but actually he may end up striking her in bed. Well before this disorder was recognized in humans, it was observed in cats with damages in their brainstem.

Do we really need to sleep? Yes, anyone who lives for months without sleep will probably die. The sleep-memory connection has long been established—more-recent work indicates that the brain strengthens different kinds of memories during different stages of sleep. Sleep, however, does not serve just a single purpose. Rather, it appears to be needed for a whole host of physiological processes, including immune response, hormone balance, emotional health, and learning ability. Don't skimp on sleep; it's bad for your health.

4.6 GENERAL ANESTHESIA AND COMA

General anesthesia is a medically induced coma that generally acts through chemical interaction with various neurotransmitter systems. Again, we suspect that one important drug effect involves shifts of brain state along the local-global spectrum of dynamic patterns, although many more specific changes must be involved. The specific mechanisms are not well-understood, but functions in the cerebral cortex, thalamus, and brainstem may be altered. Prior to an operation, the patient is evaluated to determine the best combination of drugs and dosages that will provide a safe and effective outcome. Key factors in such evaluation include age, body mass index, surgical history, and current medications. To cite one example, a patient who regularly drinks a lot of alcohol might require a heavier dose of the anesthesia drug to avoid waking up during the operation or suffering dangerously high blood pressure.

Anesthesia depth has a rating system similar to sleep stages. Stage 1 is induction, the period between the initial injection or inhalation and loss of consciousness. Stage 2 occurs just after loss of consciousness, often indicated by delirious activity. Respirations and heart rate may become irregular; spastic movements, vomiting, and other undesirable events may occur. Rapidly acting drugs are used to shorten this stage as much as possible. Stage 3 is surgical anesthesia, in which skeletal muscles relax, vomiting stops, and respiratory depression occurs. Stage 4 is overdose, where too much medication has been given; this stage would be lethal without cardiovascular and respiratory support. In simple terms, if you get too little anesthesia, you may wake up with your chest open and heart exposed; too much anesthesia, and you die. To avoid such embarrassing outcomes, several kinds of monitoring are employed in the operating room, including EKG, blood pressure, blood-oxygen level, body temperature, and so forth. Sometimes EEG is also recorded since it provides rapid responses to changes in brain function, especially when caused by low brain oxygen.

Coma is an unconscious state in which a person cannot be awakened or respond to stimuli; he or she lacks a normal wake-sleep cycle and does not produce voluntary actions. The comatose person exhibits no wakefulness and is unable to feel, speak, hear, or move. Generally, for a patient to maintain consciousness, two important brain structures must function normally—the cerebral cortex and the *reticular activating system*, a group of neural structures located in the brainstem. Most comas involve malfunctions of neurotransmitter systems. Comas may be caused by a variety of conditions. The largest number of cases results from drug poisoning; the second most common cause is lack of oxygen, often due to cardiac arrest. Since drug poisoning is a frequent cause, hospital emergency room personnel first test comatose patients by observing pupil size and eye movement. Other causes of comas are diseases of the central nervous system, strokes, and head trauma, the latter often due to falls or auto accidents. Comas may last for several days or, in more-severe cases, weeks; some have lasted years. After this time, some patients come out of the coma gradually, some progress to a vegetative state, and others die. Patients who have entered a vegetative state may go on to regain a degree of awareness, or they may remain in a vegetative state for years or even decades. The outcome for coma and vegetative state depends on the cause, location, and extent of neurological damage.[5]

4.7 EPILEPSY: A COLLECTION OF LOCAL-GLOBAL DISEASES

A disease that dramatically reveals the close relationship between brain dynamic patterns and healthy brain function is epilepsy. Actually, we may more accurately adopt the plural "epilepsies" to distinguish between the different symptoms and medical treatments associated with different parts of the local-global spectrum of epileptic activity. The diseases range from the *focal epilepsies*, which are often treated by surgical removal of the offending brain tissue, to global phenomena that call for anticonvulsive drugs or electrical stimulation.

Ancient writings on epileptic seizures go back as far as four thousand years, nearly as far back as evidence of the first civilizations. The Babylonians and ancient Greeks both attributed seizures to possession by evil spirits; the Greeks also associated seizures with genius and the divine. But in most cultures, epilepsy sufferers have been disgraced, shunned, and even imprisoned. Epilepsy-related stigma is still commonly experienced around the world, affecting the victims and their families economically, socially, and culturally.

Epilepsy is a disease affecting about one percent of the world population. Some cases are lifelong, but others improve to a point that medication is no longer needed. As represented symbolically in figure 4-4, epileptic seizures are essentially "storms in the brain" that can vary from brief and almost undetectable episodes to long convulsion periods. The causes of epilepsy are mostly unknown, although excessive excitatory neural activity and synchronization are clearly indicated by EEG recordings, as discussed in chapter 6. In simple terms, too many neurons are active at the same time; consciousness evidently requires brain patterns with many inactive neurons. Genetics is involved—studies of identical twins indicate that if one twin is affected, there is an even chance that the other twin will also be affected. In non-identical twins, the risk is about 15 percent. Some cases of epilepsy are more associated with brain injury, tumor, stroke, or substance abuse. The process of tornado formation, which also involves local (bottom-up) and global (top-down) contributions of weather patterns, may provide a plausible metaphor for seizure production.

Generalized convulsive seizures involve both brain hemispheres from the onset. Partial seizures involve only one hemisphere at the start but may later progress to generalized seizures. The nonconvulsive events characteristic of epilepsy include *absence seizures*, states of decreased consciousness that usually last only a short time. For example, a six-year-old child in the classroom may appear to "blank out" for ten seconds or so but later have no memory that anything unusual happened. Absence seizures may occur only rarely, or they may occur many times in a single day. In more than half the cases, the seizures stop when the child reaches teenage years, presumably due to normal brain development.

The first choice of treatment for epilepsy is anticonvulsant medication, possibly for the patient's entire life. The drug of choice is based on seizure type, epilepsy syndrome, other medications used, other health problems, and the person's age and lifestyle. About 30 percent of people continue to have seizures despite anticonvulsant treatment; epilepsy surgery may be an option for such patients. Full control of seizures may be achieved in 60–70 percent of surgery cases, in which selected parts of the neocortex are removed. Such tissue is typically identified with recordings directly on the cortical surface that employ implanted electrodes (ECoG).

Despite treatments with antiepileptic drugs, about half of patients continue to experience seizures or severe side effects from the drugs. Epilepsy surgery is a viable option only when seizures originate in small, known tissue masses that can be safely removed. Location of such seizure foci requires patients with implanted electrodes to be monitored in the hospital over perhaps several days while physicians wait for a seizure to occur. Another approach, which is currently effective in many cases, involves brain electrical stimulation, a fascinating and developing field with enormous potential for future medical advancement in epilepsy, stroke, Parkinson's disease, coma, and more.

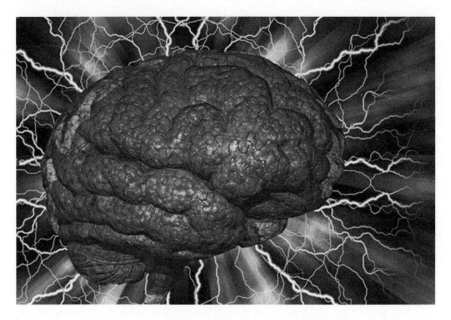

Fig. 4-4. © Can Stock Photo Inc. / rolffimages.

One method used to control seizures is *vagus nerve stimulation*. In this medical approach, brain stimulation is accomplished indirectly by implanting a pacemaker-like device that electrically stimulates the vagus nerve, one of twelve cranial nerves that conduct signals (action potentials) between the brain and other body locations. The stimulator, about the size of a silver dollar, is implanted under the skin in the upper chest, and a connecting wire is placed under the skin and connected to the vagus nerve. The stimulator's computer is programmed to generate pulses of electricity at regular intervals. For example, the device may stimulate the nerve for 30 seconds every five minutes. The settings on the stimulator are adjustable, and the electrical current is gradually increased as the patient's tolerance increases. The patient is also given a handheld magnet, which, when positioned near the stimulator, can generate an immediate electric current to stop a seizure in progress or reduce its severity. In other words, the patient is provided with direct control over his own brain activity, just by moving the magnet next to the stimulator.

Yes, brain electrical stimulation really works, sometimes spectacularly, but the burning question of just how it works remains mostly unanswered. Brain cells communicate by sending electrical signals in an orderly pattern. In epilepsy, this pattern is sometimes disrupted due either to an injury or the person's genetic makeup, causing neurons to emit signals in an uncontrolled fashion. This creates overexcitement, the so-called brain storms or seizures. The vagus nerve provides an important pathway to the brain; by stimulating this nerve, electric current is apparently spread into a wide area of the brain, disrupting the abnormal brain patterns responsible for seizures, a top-down influence. A supplemental idea is that stimulation of the vagus nerve causes the release of neurotransmitters that decrease seizure activity. These are obviously overlapping concepts since electrical events cause chemical events and vice versa by means of the bottom-up and top-down interactions, which were outlined in chapter 3.

An alternate approach to brain stimulation, partly developed by one of my former students, Brett Wingeier, employs a matchbox-sized device implanted in the skull.[6] The device records EEG and contains sophisticated software to recognize the EEG "signature" of a seizure onset. The device then delivers a mild electric stimulus to areas near the seizure focus, attempting to suppress the seizure. Brett has since formed his own company (with two partners), Halo Neuroscience, Inc., which delivers brain electrical stimulation to normal brains with the aim of improving motor function. These are just two of many new practical examples of brain-computer interfaces, including devices that allow *locked-in patients* to communicate. Michael Crichton, the medical doctor better known as a fiction writer, anticipated the epilepsy application in his 1972 novel *The Terminal Man*.[7]

4.8 LOCKED-IN SYNDROME

A very troubling ethical dilemma occurs with *locked-in syndrome*. The locked-in patient lives in a state of severe and permanent paralysis. Patients fully locked in suffer from a complete and irreversible loss of motor functions, making it impossible for them to acknowledge or communicate with others even though they may maintain near-normal levels of consciousness. They often understand what others are saying and doing, but they cannot respond. The locked-in state is very much like the state caused by curare poisoning, the paralysis of all voluntary controlled skeletal muscles.

The locked-in state can occur as a result of a stroke, a head injury, multiple sclerosis, or other causes, including snake bites that inject neurotoxins. A stroke in the brainstem may cause complete paralysis by blocking all signals from the brain to muscles but leave the cerebral cortex (and consciousness) fully intact. End-stage amyotrophic lateral sclerosis (ALS) is also a common cause of the locked-in state. This condition is also known as Lou Gehrig's disease, after the famous baseball player of the 1930s. ALS afflicts roughly one in a thousand people with no known cause. Today's most famous ALS patient is renowned theoretical physicist Stephen Hawking. Although he is unable to move or speak, he still communicates by moving his little finger to choose words for his speech synthesizer. He has managed to produce numerous scientific papers and books on cosmology, over more than forty years after diagnosis. While we applaud Hawking's astonishing scientific and personal accomplishments under such severe conditions, we should not fool ourselves that other ALS patients are likely to come close to duplicating his success. Hawking has survived his disease much longer than most other ALS patients, who typically die within two to five years of diagnosis.

We can only speculate on the appalling experiences of patients in the locked-in state. Some patients appear to have normal consciousness. Others may be in a state of stupor, appearing halfway between coma and consciousness, only roused occasionally by appropriate stimulation. They are unable to tell their stories, which possibly contain rich mental experiences. Often the patient's wishes for life support are unknown. Patients are totally paralyzed; they can think, feel, and know, but they cannot express themselves in any manner. Only diagnostic tests like EEG and fMRI reliably distinguish between the fully locked-in state and the vegetative state. The locked-in patients are essentially *anti-zombies*. Whereas philosophical zombies (p-zombies) simulate human consciousness that they lack, locked-in patients simulate a lack of the consciousness that they actually possess. The phrases "living hell" or "buried alive" come immediately to my mind when I consider their plight. Current research in creating brain-computer interfaces employs the patient's EEG as a means of direct communication to the outside world. The plight of locked-in patients provides strong motivation for such research.

Perhaps the second most remarkable story of a semi-locked-in patient is that of Jean Bauby, once editor-in-chief of a French fashion magazine. At age forty-two, Bauby suffered a massive brainstem stroke that left him paralyzed and speechless, able to move only one muscle, his left eyelid. Yet his mind remained active and alert. By signaling with his eyelid, he dictated an entire book titled *The Diving Bell and the Butterfly*,[8] blinking to indicate individual letters, a total of about two hundred thousand blinks to "narrate" the book. The book led to a movie with the same title. Bauby's book and movie tell about life in the hospital, flights of fancy, and meals he can eat only in his imagination. His story is also concerned with his ability to invent a new inner life for himself while trapped in the proverbial diving bell, providing us with valuable insight into a unique kind of consciousness. Bauby died five days after the book's publication.

At least several thousand patients in the United States alone now exist in the irreversible conditions of deep coma, vegetative state, minimally conscious state, or locked-in state. Most are unable to communicate to family and caregivers whether they would choose to continue living in such dire circumstances or assert their right to die. Since all of us are potential candidates for these conditions because of automobile accidents, stroke, or diseases like ALS, recording our wishes in advance is essential. Such instructions to our designated agents or family members must consider several issues, including religion, worldview, age, financial capacity, and the law's limitations. In the case of locked-in patients, the potential anguish of being trapped in the proverbial diving bell or buried alive must be considered. Ideally everyone "at risk" (meaning *everyone*) will have created a personal medical durable power of attorney while still able.

Research on brain-computer interfaces is also directed to the task of establishing communication with locked-in patients.[9] Patients in the fully locked-in state are unable to communicate with others even though they may possess near-normal levels of consciousness. They can understand what others are saying and doing, but they cannot respond either verbally or by moving body parts; in extreme cases, even eye blinks are not possible. For such patients, EEG (scalp) or ECoG (from implanted cortical electrodes) are all that remain for communication. A patient viewing a computer monitor can "learn," by some unknown process, to modify his EEG to move a cursor toward target letters on the screen, eventually spelling out entire words. While this process is quite tedious, it represents a vast improvement over the cruel state of complete isolation.

4.9 BRAIN INJURIES AND LANGUAGE

One of the oldest and most rewarding approaches to brain science employs observations of the detailed symptoms caused by various brain injuries and diseases. The resulting physical and mental impairments can be amazingly focused. For example, *aphasia* refers to a group of language disorders caused by brain damage. Such damage may be due to a stroke, a tumor, an automobile accident, a gunshot wound, and so forth. Two major cortical language centers have been studied for years. One is *Wernicke's area* (shown in figure 4-5), which is located near the intersection of the temporal and parietal lobes, most often in the left hemisphere. Someone suffering damage to this region may produce natural-sounding speech and normal syntax but employ entirely meaningless words. He or she may also be unable to understand either written of spoken language. This condition is known as *Wernicke's aphasia*.

The second major language region is located near the intersection of the temporal and frontal lobes and controls speech. With damage to this region, the patient may know just what he wants to say but just can't get it out, a malady called *Broca's aphasia*. Most of us have had this kind of experience: "I'm trying hard to remember the name of the actor who played in *Jurassic Park*. I can picture his face clearly and even know some of the other movies that he played in. His name is on the tip of my tongue, but I just can't access it." Studies of various language deficits could fill many books. These two well-known aphasia cases are chosen here in the context of the local-global spectrum of dynamic behavior. While the selective importance of local brain regions is emphasized by these classical studies, more-recent neuroimaging investigations suggest that the mental and physical functions associated with these areas are more widely distributed in the cortex than previously thought.

All information relayed in both directions between the body and cerebrum must traverse the brainstem. The ascending pathways from the body to the brain include sensory pathways as well as tracts associated with pain, pressure, and temperature sensations. The brainstem is involved with control of heart function, respiration, pain, and alertness and consciousness. Thus, brainstem damage is a very serious and often life-threatening condition. Strokes in the brainstem can result in widespread muscle paralysis while leaving the cortex intact. The result can be a locked-in patient, one who has near-normal consciousness but cannot communicate.

The complex inter-relationships between language, brain disorders, and consciousness provide a long and fascinating story, but here we limit this material to few examples focused on local-global brain dynamic connections to language. Words are collections of syllables associated with ideas, yet most words are far more complex than just literal meanings; they also carry emotional baggage as a result of their broader context. Brains do indeed contain local Wernicke's and Broca's language centers; however, actual language production is influenced to varying degrees by other parts of the brain, as well as the external environment.

For some words, this emotional connotation is so intense that even in the United States, where freedom of speech is a fundamental right, the use of these words can be officially or unofficially banned. The consequences of using them in "inappropriate" contexts can range from censorship and

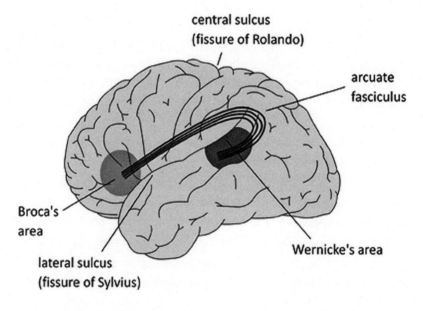

fines to ostracism and possibly even loss of one's own TV show. In some repressive countries, words can even get you killed. In spite of these dangers, swear words, taboo phrases, and other forms of *curses* persist across societies and throughout history—a product of culture, language, and the brain itself. Sometimes I wonder just who ultimately decides which words are good or bad—is some super-secret government agency involved? The distinction between good and bad words varies substantially from one generation to the next. In 1939, actor Clark Gable created a minor sensation by uttering the famous words, "Frankly my dear, I don't give a damn" in the famous movie *Gone with the Wind*. The linguistic innocence of those days is now hard to fathom.

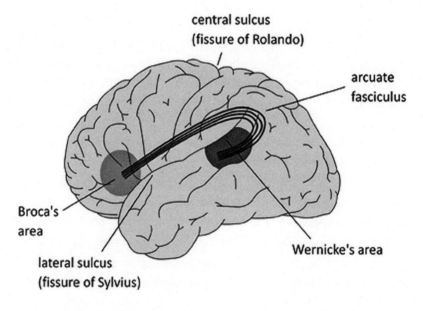

Fig. 4-5. *Wikipedia* Creative Commons, author: Peter Hagoort.
Licensed under CC BY 3.0.

A well-known neuropsychiatric disorder is *Tourette's syndrome*, which is inherited and apparently not caused by localized tissue damage. Tourette's is characterized by sudden repetitive movements (physical tics) and involuntary sounds produced through the nose, mouth, or throat (phonic tics). Tourette's is often viewed as a bizarre syndrome associated with blurting out obscene words or inappropriate and derogatory remarks. However, this odd symptom actually occurs in only a minority of Tourette's patients, but it can cause hurtful repercussions with the friends and family of those afflicted.

In an episode of the TV series *Curb Your Enthusiasm*, producer and lead actor Larry David hires a chef for his new restaurant. Larry later discovers that his new chef suffers from Tourette's syndrome and is likely to start swearing for no reason at any time. Any such outburst could cause serious problems because the restaurant's chef station is located in close contact to the customers. On opening night, the restaurant is closely packed with Larry's friends, family, investors, and more. Of course, the inevitable occurs; the chef lets out a thundering profanity streak that stuns the customers; the restaurant then goes deadly quiet. Larry is in a panic but remembers seeing news programs where friends of bald cancer patients undergoing chemotherapy show solidarity and friendship by shaving their own heads. Larry's brain lights up—he responds with his own string of obscenity in solidarity with his chef. The customers soon catch on, and the entire restaurant is transformed into an "obscene orchestra" supporting the chef. This episode is not only humorous but provides a creative heart-warming gesture supporting the handicapped.

Aphasia can also feature excessive swearing. The specifics of a particular aphasia depend on the location and severity of the damage; in general, though, aphasic individuals have problems with speech, listening, reading, and writing. In the most severe case, global aphasia, speech is mostly absent. Yet many of these patients are able use swear words easily with the proper pronunciation. Individuals with aphasia have damage to the normal language areas. The fact that they are able to swear suggests that swearing involves more extensive areas and is processed differently than other parts of language. Again brains operate through a combination of local and global processes.

The medical symptoms associated with brain damage depend strongly on both the detailed location and size of the damaged region. In the context of the broad range of brain diseases, we may reasonably postulate that some symptoms are caused directly by the destruction of local network structures, possibly the critical nodes of extended networks. On the other hand, other symptoms, especially those with strong mental aspects, may be due to more global deficits. These global effects may be caused partly by local systems failing to act normally or by acting bottom-up to disrupt healthy large-scale dynamic patterns. One way for this process to occur is the destruction of critical network hubs as depicted symbolically in figure 4-6.

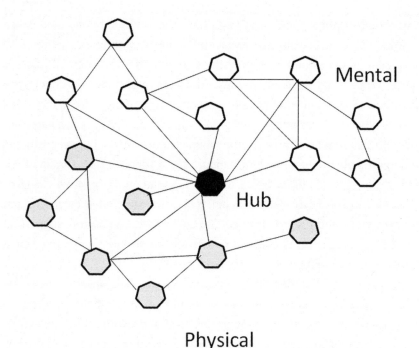

Fig. 4-6.

The gray nodes are symbolic representations of neuron groups mostly associated with physical activity, whereas the white nodes are involved mainly with mental activity. These brain functions are bound together by an imagined major hub shown in black. We call this a "hub" because it has more than twice as many connections as any other node. This is a hub in the same sense that Atlanta, Georgia, is a hub for Delta Airlines. Damage to a non-hub node may not cause major functional deficits in the brain;

however, hub destruction might be catastrophic for integrated physical-mental processes like singing while playing a guitar. Likely candidates for this process involve any damage to local neuromodulator systems that fix healthy brains somewhere near the midrange of the local-global spectrum of dynamic pattern behavior.

4.10 SUMMARY

We began this chapter with the obvious question—just what is consciousness? While conscious experience is the most familiar aspect of our lives, it is also the most mysterious. All proposed definitions of consciousness are circular, suggesting that consciousness may be a fundamental property in the sense that electric charge is fundamental. One plausible definition of consciousness is *the state of awareness of an external environment or of something within oneself; the ability to experience or feel*. Despite this difficulty with definition, consciousness is now considered to be a legitimate research topic in psychology and neuroscience. Determining the neural and psychological correlates of consciousness, the so-called easy problem, is the primary focus of this work. The hard problem, the origin of consciousness itself, may be approached cautiously through studies of the easy problem. Finding such consciousness correlates will not, in and of itself, solve the hard problem; however, the "easy" information serves to mold our ideas about the hard problem.

The essay "What Is It Like to Be a Bat?" presents an influential thought experiment that argues against several versions of materialism and reductionism. Materialist positions must take into account known physical laws, including the laws of quantum mechanics. This field of modern physics reveals the existence of a hidden of shadow world of some kind that influences our real world; such shadow world may or may not have anything to do with consciousness.

Patterns provide important aspects of brain activity. Such dynamic patterns are closely correlated with human behavior and conscious activity—patterns of electric fields, blood-oxygen level, metabolic activity, and more. The label "dynamic pattern" is defined broadly to encompass brain measurements that are currently in use, plus different kinds of patterns that may be measured in the future.

Emphasis is here placed on analogue and metaphorical networks, consistent with our goal of dealing with basic ideas rather than getting bogged down in technical details. However, our analogue networks mirror a number of genuine brain networks. The structure of the cerebral cortex consists of complex parts that occur at multiple spatial scales—the cortical columns and so forth. This fractal-like structure suggests that the various kinds of spatial patterns measured by neuroscientists should also be strongly scale-dependent. Cross-scale interactions may be essential to the production of consciousness. Various mental states are summarized to demonstrate these ideas.

Different stages of sleep correspond to different levels of consciousness—stage 1 is the drowsy state; stages 2 and 3 are deeper. The rapid eye movement (REM) stage is closely associated with dreaming. Sleep stages and waking conditions are controlled by neuromodulators, and the largest contributions come from cells in the brainstem. Brain and mind cycle through the various sleep stages every night. Distinct stages are clearly revealed by recordings of eye movements, muscle tension, and the electroencephalogram (EEG).

General anesthesia is a medically induced coma that generally acts through chemical interaction with various neurotransmitter systems, but the specific mechanisms are not well-understood.

Coma is an unconscious state in which a person cannot be awakened or respond to stimuli. He or she lacks a normal wake-sleep cycle and does not produce voluntary actions. Comatose persons exhibit no wakefulness and are unable to feel, speak, hear, or move. Most comas involve malfunctions of neurotransmitter systems.

Epilepsy is a disease exhibiting especially strong relationships between brain dynamic patterns and healthy brain function. Seizure patterns recorded with EEG have very large amplitudes associated with too many excitatory neurons, extreme synchronization, or both. EEG waveforms are highly abnormal and provide unambiguous seizure signatures. Different symptoms and medical treatments are associated with different parts of the local-global spectrum of brain dynamic activity. Various aspects range from the *focal epilepsies*, which are often treated by surgical removal of the offending brain tissue, to global phenomena that require anticonvulsive drugs or electrical stimulation.

The medical condition locked-in syndrome presents a troubling ethical dilemma. Locked-in patients live in states of severe and often permanent paralysis. They suffer from complete loss of motor functions, making it impossible for them to acknowledge or communicate with others even though they may maintain near-normal levels of consciousness. They often understand what others are saying and doing, but they cannot respond.

Scientific observations of symptoms caused by various brain injuries and diseases provide a rich contribution to brain science generally and to the study of consciousness in particular. The resulting physical and mental impairments are often quite focused. *Aphasia* refers to a group of language disorders caused by brain damage like strokes, tumors, accidents, and so forth. Two major cortical speech centers, *Wernicke's area* and *Broca's area*, have been studied for many years. Damage to Wernicke's area involves loss of understanding of language; the patient may produce natural-sounding and normal syntax, but he or she will employ entirely meaningless words. By contrast, with damage to Broca's area, the patient may retain full understanding; he may know just what he wants to say but is unable to say it. These two well-known aphasia syndromes are chosen here in the context of the local-global spectrum of dynamic behavior. While the selective importance of local brain regions is emphasized by classical studies, more-recent neuroimaging investigations suggest that the mental and physical functions associated with these areas are more widely distributed in the cortex than previously believed. In the following chapters, we will look more closely into the nature of these distributed dynamic patterns that are so essential to healthy mental functions.

Chapter 5

SIGNATURES OF CONSCIOUSNESS

5.1 WHAT IS TIME?

The issues of consciousness and free will are intimately connected to the passage of time. Like consciousness, time is a familiar entity, but time is actually quite mysterious when considered in scientific depth.[1] Physicists of the nineteenth century pictured the universe as composed of particles of matter acted on by forces causing them to move through a static space. I suspect that most people, including most of today's scientists, still see things pretty much this way. This view is part and parcel of our intuitive heritage, which evolved over several hundred thousand years. If this imagined, but false, classical universe is analogous to a vast cosmic play, the particles are the actors, and space and time make up the fixed stage. In the nineteenth century, the job of classical scientists was seen as discovering details of the cosmic play's plot. This logically intuitive idea was dramatically debunked by Einstein's special theory of relativity in 1905, his general theory published in 1915, and developments in quantum mechanics over the first thirty years or so of the nineteenth century. However, a whole generation of scientists had to die off before the new ideas were fully accepted even within the physics community. One wonders if the history of consciousness science might follow an analogous path. To cite one prominent question intimately involved with time—will our apparent ability to carry out freely chosen future actions be exposed as just an illusion? Or is such *free will* an essential aspect of consciousness?

Einstein's greatest accomplishment was to show that the "god's truth" about the physical universe is truly shocking—the cosmic play's distinction between cast and stage is false. It's just one of those inventions of human minds that probably aided in our survival as a species. If a tiger is chasing you through a jungle space, Einstein cannot help you. But in reality, space and time are part of the cosmic play's cast; space and time are physical things and, like matter and energy, are subject to known physical laws. In modern physics, our space-time universe limits information transfer to the speed of light; space and time are severely distorted by massive bodies; space is rapidly expanding; quantum fields appear and disappear out of nowhere, and more. In the words of physicist John Wheeler—"time is clothed in a different garment for each role it plays in our thinking."[2] In order to understand consciousness, perhaps we must first understand time. Unfortunately, nobody seems to understand time.

"OK," you say, "But what do these abstract, esoteric ideas have to do with the familiar consciousness that is somehow 'embedded' in our universe?" My short answer is this: maybe not much, but maybe a whole lot. One thing we do know with near certainty—"life" as we know it developed within a space-time universe that is life-friendly. Is consciousness an inevitable consequence of life, or is something additional involved, call it the "C-factor," which might be present in some imagined universes but not in others? We will return to such notions in the more-speculative later chapters concerned with ontological issues that overlap both physics and consciousness, but for now, we stick with the classical concept of time, which is closely interrelated with the challenge of consciousness.

5.2 CONSCIOUSNESS TAKES TIME

In chapter 2, we cited fascinating earlier studies of consciousness carried out by physiologist Benjamin Libet and his colleagues.[3] Their research focused on differences between the brain's registration of a stimulus and conscious awareness of the same stimulus. Why are these two events not identical? The fact that they differ makes these studies quite interesting because *unconscious mental functions act over different time intervals than conscious mental functions.*

One of Libet's experiments involved brain-surface electrodes implanted in epilepsy patients. Direct electrical stimulation of the *somatosensory cortex* (the brain region reacting to sensory input) in waking patients elicits conscious sensations like tingling feelings. These feelings are not perceived as originating in the brain; rather, they seem to come from specific body locations determined by neural pathways. Stimulation near the top of the brain near the midline (the motor cortex) produces tingling in the leg; stimulation points on the side of the brain produce feelings from parts of the face, and so forth. Libet's stimuli consisted of trains of electric pulses, and the essential stimulus feature was found to be the duration of the stimulus. Only when stimuli were turned on for about half a second (500 milliseconds) or more did conscious sensations occur. Experiments were also carried out by stimulating deep brain regions, with similar results.

Why did these scientific studies require brain stimulation; why not just stimulate the skin or flash a visual image? Such short stimuli register in our consciousness because skin or visual stimulation passes through multiple processing stages in both the peripheral system and deep brain, resulting in input to the cerebral cortex that is spread out over time. Recording electrodes implanted in the somatosensory cortex begin to show a response from stimuli after perhaps 5 to 50 milliseconds, depending on stimulus locations. Shorter latencies occur when the stimulus site is on the head; stimulation of foot areas takes much longer. These delays result mostly from signal propagation (*action potentials*) delays along peripheral nerves of differing lengths. The first electrical responses in the cortex, called *primary evoked potentials*, do not by themselves indicate conscious awareness; electrical responses must persist for much longer times for awareness to occur.

Libet's studies suggest that consciousness of an event is only associated with electric field (or electric potential) patterns in neural tissue that persist for at least 300 to 500 milliseconds. To check this idea, Libet and others have carried out several versions of *backward masking studies*. The general experimental setup provides two distinct stimuli. If the second stimulus is applied between about 200 milliseconds and 500 milliseconds after the first stimulus, it can "mask" conscious awareness of the first stimulus, meaning the subject will report no knowledge of the first stimulus.

A child brain's judgment of time is due to actions in widely distributed brain tissue that develop over at least several years. Two- and three-year-old children's understanding of time is mainly limited to "now" and "not now." In contrast, five- and six-year-olds grasp the ideas of past, present, and future, thereby satisfying plausible conditions for the *extended consciousness* suggested by Antonio Damasio (discussed in chapter 3). Time is closely related to the consciousness issue in many ways, impacting both the easy and hard problems as outlined below.

- Our sense of "now" is distributed over about one-half second in the past; it's the "remembered present."
- Conscious awareness of an external event requires at least several hundred milliseconds to develop, involving many passes of signals back and forth between multiple brain regions. Consciousness is believed to be associated by the total dynamic activity of distributed cortical networks rather than any one network node or by any one instant in time. This distributed brain system is sometimes labeled *the global workspace*.[4]
- Consciousness requires special kinds of focused synchrony between activities in distant brain regions; it's a bit like an orchestra producing beautiful music.
- Unconscious mental functions, including reflexes and pre-conscious processes, operate over faster time intervals than conscious functions do.
- The proper timing of internal signals is essential to healthy brain function. Learning is believed to occur when multiple action potentials arrive nearly simultaneously at a postsynaptic neuron, thereby strengthening the connection. *Neurons that fire together, wire together.*
- A broad range of psychiatric and developmental disorders involves myelin defects in white matter. Myelin controls action potential speed and the "synchrony" (matched time delays) of impulse traffic between cortical regions.
- Different mental states are closely tied to EEG *brain rhythms*, the time-dependent oscillations of electric field patterns.

- Remote networks may interact selectively through *resonance*, the general tendency of some systems to respond only to a narrow range of frequencies, as in your cell phone antenna—resonance is closely related to synchrony issues.

It's fair to say that all these entries, excepting the last one involving resonance phenomena, are well established. In chapter 8 we will look into the brain-resonance question in some depth, but for now, note that resonance occurs nearly everywhere—in bridges, TV tuners, fMRI technology, chemical bonds, planetary orbits, and musical instruments. Examples from neuroscience include vision and hearing—for example, detection of light by cells in the retina is basically a quantum-resonance phenomenon manifested at the atomic-molecular scale, although neuroscientists seldom describe it in these terms. Later, I will propose the addition of neural networks to this list of resonant systems.

5.3 PRE-CONSCIOUS AND CONSCIOUS STATES

Much of this section relies on the writings of neuroscientist Stanislas Dehaene, encompassing his own fascinating research plus discussions of similar studies by colleagues. These methods manipulate conscious and unconscious processes with clever experimental designs using visual images.[5] Their basic idea is this—by employing masking strategies, forced inattention, or very brief image exposure, any picture presented to a subject can be made to vanish from consciousness. Or it can be placed at an experimental design threshold so that it is perceived by subjects only half the time. In this manner, consciousness is the only manipulated variable—the subject reports seeing the image in some cases but not in others.

Even very small changes in experimental design can cause subjects to see or not see some image on a screen; for example, a word may be flashed so quickly, perhaps for 50 milliseconds, that viewers fail to notice it about 50 percent of the time. The perceived image, the one that makes it into the experimental subject's awareness, and the lost image, which vanishes into the depths of the unconscious, may differ only a little on the input side. But, within the brain, this difference is amplified or otherwise modified when the subject becomes aware of one image but not the other. Illuminating these brain-pattern modifications is the focus of this section.

Research suggests that massive unconscious processing occurs beneath the surface of our conscious awareness. Several different measures of brain activity, employing fMRI and EEG methods, undergo quick changes when a person becomes consciously aware of a face, a word, a number, or a sound. Once dynamic patterns carrying information reach the level of consciousness, they are no longer constrained by sensory pathways; they can then develop in alternate paths, apparently as manifestations of *free will*. But just where does such free will originate? Why, the mind, of course! But, if this circular logic leaves you cold, you are not alone.

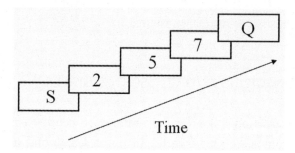

Fig. 5-1.

Studies of conscious and unconscious processing often involve subjects viewing a series of images presented on a computer screen, as indicated in figure 5-1. Most of the symbols in this example are digits, but some are letters that the subjects are told to remember. The first letter, *S*, is easily remembered. If the second letter, *Q*, is presented more than a half second after the first, it is also remembered. By contrast, if the interval between the two letters is too short, the second letter will often be missed. A subject may be quite surprised to learn that two letters were presented.

In other tests, the subjects may be asked to guess things about pictures shown to them, such as their location in space. They are often able to score well above chance level, showing that the pictures entered the unconscious, but this information never rose to the conscious level. In general, nearly any visual image that is flashed for a very short time or is masked by a second image may enter the unconscious. We may call this part of the unconscious the *pre-conscious*, because it later influences conscious decision making in a measureable manner. Such subliminal processing of visual images is known as *blindsight*. In related experiments performed by several research groups, patients who are blind because of damage to their primary visual cortex (where signals arrive after relay through the thalamus) assure us that they can see nothing at all. Yet, when asked to guess about an assortment of pictures shown to them, they are often able to give the correct answers; the unconscious reveals its secrets in this manner.

Not long ago conventional wisdom associated the cortex with conscious activity and subcortical tissue or networks with unconscious processes. The cortex, considered to be the most advanced part of the mammalian brain, is largely responsible for planning, speaking, and abstract reasoning. It was thought that whatever information reached the cortex would probably be conscious. Another old idea was that some networks are conscious while others are not. Neither of these ideas seems to have survived the closer scrutiny of modern brain imaging. A more up-to-date view is that nearly all of the brain's regions can participate in both conscious and unconscious activity. Later we will argue that conscious and unconscious networks may occupy overlapping locations in the brain.

EEG, including evoked or event-related potentials (ERPs), and fMRI reveal how brain activity develops and spreads as information supplied by the senses gains conscious access. ERPs are scalp electric potentials, caused by visual, auditory, or other stimuli, a subcategory of EEG. By contrasting *conscious-related* with *unconscious-related* measures of various kinds of brain activity, scientists find evidence that the stimulus was consciously perceived. Dehaene lists four "signatures of consciousness" that he and others have identified with this approach:[6]

- Although a subliminal stimulus can propagate deeply into the cortex, fMRI activity is "strongly amplified" when the threshold for awareness is crossed. Such activity then invades many additional regions, leading to a sudden increase in blood-oxygen level in parietal and prefrontal cortical circuits.

- Conscious access appears as a late, slow waveform of the ERP (event-related potential) called the *P300*, recorded on the scalp approximately 300 milliseconds after the stimulus.

- Electrodes placed inside the human brain reveal a sudden burst of high-frequency oscillations in the EEG following the usual event-related potential (P300) waveform. Such rhythms, often in the 40 Hz range, are called *gamma activity*.

- These gamma oscillations are synchronized between some brain regions. Such correlated activity appears to be an important factor in conscious perception.

Our brains are hosts to a whole set of unconscious systems that constantly monitor the world and assign values to this information; these values guide our attention and shape our thinking. The prefrontal cortex seems to play a key role in processing new information and integrating it into unfolding plans. Dehaene suggests that this process seems to involve a division of labor analogous to an army of workers in a basement that do the dirty work of shifting through massive amounts of raw data. "Executives" at the top then make the final decisions. But just what are these mysterious "executives" and how do they arrive at decisions?

Whenever the conscious mind is occupied, all other information must apparently wait its turn in "storage" of some sort. Parts of the unconscious may be viewed as incompletely formed consciousness, that is, pre-conscious processes from which consciousness may emerge later. Other parts of our unconscious remain forever hidden from awareness but may exert important influences on our conscious mind—they affect our choices to act in certain ways. Interactions occur in both directions; the conscious mind may influence the unconscious, and vice versa.

Initiation and guidance of voluntary acts by the unconscious is a common occurrence familiar to anyone who has ever played musical instruments or participated in sports. The complex responses required of our muscles by these activities are much more involved than simple reflexes. A basketball or tennis player makes split-second adjustments according to his location, velocity, and body angle. Conscious planning of very quick action in many activities is typically detrimental to performance; it is best if our painfully slow conscious mind relinquishes control to the much-quicker unconscious.

5.4 THE CHRISTMAS-TREE BRAIN

The scientific work by Dehaene and his colleagues engaged in similar research is fascinating; however, in this section we examine the basic question of just how much genuine "understanding" of consciousness or consciousness correlates has actually been achieved, especially in the context of the provocative label "signatures of consciousness."

In order to present a lucid nontechnical discussion of the issues involved, we again visit the Christmas-tree brain metaphor; however, this time our simple tree will be upgraded to a *magic tree* so as to provide a more-genuine analogue. Our fanciful tree holds perhaps fifty to a hundred large light bulbs like the one on the left side of figure 5-2, analogous to the regional scale of cortical tissue of several centimeters. The brightness level of some of the burning lights is analogous to electric-field intensity (amplitude), as revealed by EEG or another measure. The brightness of other lights is analogous to the local blood-oxygen level measured with fMRI. In any tree location, these distinct measures of so-called local brain "activity" may or may not agree with one another. They might also "match" based on some criteria, but disagree if compared in other ways.

Upon looking closely inside the bulb on the left side, we discover that the large bulbs actually consist of many smaller light sources; you might picture about a hundred tightly packed midsized lights, analogous to *macrocolumns* at the millimeter scale. But we don't stop our magnification exercise there. When we look closely inside one of the millimeter-sized lights, we find that each one is actually composed of one hundred or so tiny lights (shown at the upper right); these lights are then analogous to 0.3 mm scale of *cortico-cortical columns*. Again we look inside sources at this tiny scale to discover the even smaller sources (shown in the lower right corner) analogous to cortical *minicolumns*. This process of progressive magnification of "Christmas-tree-activity regions" may be continued down to the scale of *single neurons* and beyond, even down to *molecular structure*—in short, the lights form fractal-like patterns.

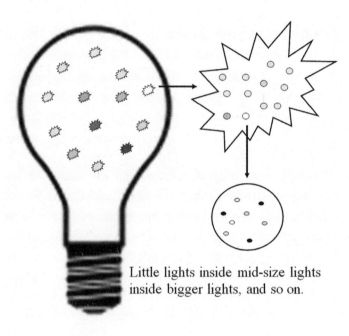

Little lights inside mid-size lights
inside bigger lights, and so on.

Fig. 5-2.

Nearly all the lights in our Christmas tree are flickering, generally with mixed multiple frequencies, providing the "dynamic" aspect of dynamic patterns. Scientists typically refer to systems consisting of multiple organizational levels as exhibiting distinct behaviors at macroscopic, mesoscopic, and microscopic scales. But this three-level language is far too limiting to accurately describe brain tissue; table 3-1 (in chapter 3) lists six spatial scales ranging from minicolumn to hemisphere, as determined by the anatomy and the different kinds interactions occurring at each scale. These six levels do not, however, cover the entire range of possible consciousness correlates.

Additional markers of consciousness, discovered at the tiny scale of single neurons, have amazed many neuroscientists, myself included. Neurons fire action potentials on a regular basis; to first approximation we may consider each brain neuron to be either "on" or "off." When turned on, a neuron is in the process of "firing" an action potential that travels along the neuron's axon and later impacts other (target) neurons. In these kinds of studies, tiny electrodes are implanted either inside or very close to single neurons, in animal brains or in human epilepsy patients. In the late 1950s, two famous scientists recorded the activity of single neurons in the primary visual cortex of anesthetized cats, and they showed how single neurons respond to very specific features of the stimulus—the vertical or horizontal direction of a line, for example.[7]

Especially in human subjects, such selective recording methods have recently yielded even more remarkable findings—some neurons can be amazingly selective when presented with visual images. Many neurons have been discovered that react only to specific pictures, places, or people. They fire only when conscious perception of the image occurs. Excitatory neurons typically fire only a few times per second. By exposing a patient to hundreds of pictures, one or two pictures may trigger a specific neuron to fire much faster. One neuron may fire only in response to a picture of Einstein, another may be a "Bill Clinton neuron," and so forth. These "face-specific neurons" do not respond to other persons.

Does this mean that if your "spouse neuron" dies, you no longer recognize him or her? No; apparently the correct interpretation of these experiments is that each specific face produces a particular dynamic pattern, possibly widely distributed, consisting of active and inactive neurons. Your brain may contain many "spouse neurons," but they may not be so easy to find in genuine experiments. A future technology that allows one to accurately locate spouse neurons might ultimately encourage various kinds of selective neuroscience-based divorces, as in some future TV ad promoting "full-frontal spousal erasure."

5.5 CHRISTMAS LIGHTS, ELECTROPHYSIOLOGY, AND FMRI

In this section, we attempt to place the remarkable findings of the last few sections in broader contexts. All scientists myself included, risk getting carried away by fascinating discoveries, perhaps fooling ourselves that some Holy Grail has been uncovered and our Nobel Prize is in the mail. In some cases of new findings, this sort of breakthrough actually occurs; in most cases, it does not.

Table 5–1. Multiple Scales of Electrophysiology.

Recording Method	Typical Resolution (mm)
Microelectrode of radius R	$\geq R$
Local field potentials (LFP)	0.1 – 1
Cortical surface (ECoG)	2 – 5
High-resolution EEG	20 – 30
Untransformed EEG	50

The science of *electrophysiology* refers to studies of electrical properties of individual cells as well as larger-scale tissue volumes. For our purposes we need not distinguish between measurements of electric fields, electric potentials, or electric currents; they are interrelated with the technical methods employed in recording. The resulting multiscale technology is outlined below and in table 5-1; the right column lists the estimated spatial resolution in millimeters for each recording method.

- *Microelectrodes* with tip sizes of about 0.001 mm typically detect the activities of single neurons when introduced into the brain of a living animal or human.
- *Local field potentials* (LFP) are obtained with larger electrodes that record the activity of multiple nearby neurons.
- *Electrocorticography* (ECoG) refers to recording from electrodes placed on the cortical surface.
- *Electroencephalography* (EEG) usually indicates recordings from

the scalp surface. Subcategories include spontaneous EEG like alpha rhythms, transient evoked potentials (EPs), event-related potentials (ERPs) like the P300, and steady-state visually evoked potentials (SSVEP).

The entry "High-resolution EEG" in table 5-1 refers to computer-based spatial filtering of raw EEG data, which will be discussed in chapter 6. With these few technical points in mind, we return to our multiscale magic Christmas-tree lights from figure 5-2. Recording EEG is very much like following the dynamic patterns of the largest light bulbs. From our god's-eye view we know that these large lights are actually composed of progressively smaller lights, but all we can measure with EEG are crude space averages over each large bulb. In spite of this severe limitation, scalp-recorded EEG has provided robust signatures of conscious states since the 1920s. We return to these large-scale brain rhythms in chapter 6, where it is shown that (historically) the most reliable signatures of consciousness in humans have been the large-scale scalp rhythms occurring at frequencies below about 15 to 20 Hz. These are the well-known *delta*, *theta*, *alpha*, and *beta* rhythms.

The coming discussion of slow scalp rhythms in chapter 6 may, at first, appear somewhat conflicted with this chapter's emphasis on fast (gamma) oscillations. Some differences between observations of "fast" versus "slow" frequencies as consciousness markers are due to differences between human and the various species of animal recordings. However, the major differences almost certainly come into play because of the different spatial scales involved in the measurements. Gamma rhythms, also known informally as "40 Hz activity," are found inside the skull, but they are very difficult to measure reliably on the scalp because of technical reasons outlined in chapter 6. If we employ the nested lights of our magic analogue Christmas tree, we will see that lights tend to flicker at mixtures of several different frequencies at all scales. However, *the actual frequencies measured depend critically on the spatial scale adopted in the experiment*. Fast frequencies are mostly missing at the scalp; intracranial recordings tend to downplay slow frequencies. This is a critical point that has often been underappreciated or even completely misunderstood by many neuroscientists. We will expand on this issue in chapters 6 and 7.

In the case of fMRI measures of blood-oxygen level, the experimental limitations are quite different from those faced in EEG and other branches of electrophysiology. The good news is that blood-oxygen levels are measured in 1 mm scale tissue voxels, roughly the size of a macrocolumn. This spatial resolution is much better than that provided by EEG, and it is as good as or better than ECoG. But the really good news is that fMRI provides three-dimensional images of the whole brain, and no electrodes need apply for implant duty.

In the case of our fMRI analogue, the light patterns produced at the 1 mm scale are depicted as the small irregular shapes inside the large bulb on the left side of figure 5-2. The brightness of these lights can be accurately measured at all these locations; however, the smaller scales shown on the right side remain "under the fMRI radar." The bad news is that fMRI has poor temporal resolution due to slow tissue blood-level responses reacting to external events. Thus, fMRI technology is quite similar to applying time-lapse photography to the flickering lights of figure 5-2. By contrast, EEG technology is like shooting a super-high-resolution (temporal-resolution) video of the largest light patterns.

High-definition videos shot to look like film take successive pictures at perhaps 24 to 60 frames per second. EEG waveforms are normally sampled at more than one hundred times per second, allowing scientists to study rhythms with frequencies up to 50 Hz or more. By contrast, fMRI may be sampled at once every few seconds. The resulting patterns are severely "smeared out" in time like EEG is "smeared out" in space; fMRI spatial resolution is perhaps fifty times better than EEG. Actual EEG temporal resolutions often depend on the trade-offs discussed in chapter 6, but temporal resolutions are often something like one hundred times better than fMRI. But, in fairness, fMRI finds patterns in the entire brain, whereas EEG mainly measures cortical activity. The main drawback of fMRI from a pattern-recognition perspective is that the "dynamic" part of "dynamic pattern" is mostly lost. Our main take-home message is that EEG and fMRI provide complementary measures of brain patterns; they measure different things at different spatial and temporal scales.

5.6 FREE WILL?

Free will is our apparent ability to choose between different courses of action. As the famous baseball player Yogi Berra once said, "When you come to a fork in the road, take it."[8] Most of us take our ability to choose our own paths for granted, but the free-will issue repeatedly surfaces in connection with consciousness. One reason that this "inconvenient" topic keeps reappearing is that it seems impossible to reconcile free will with any strong or even moderate version of determinism, in which future events are supposed to follow directly from the current state of things according to the laws of physics. In the strong deterministic view, I had no choice other than to write this book, and you are destined to finish reading it (or not) whether you "want to" or not.

Yes, more than a few intelligent philosophers and scientists actually claim that free will is just an illusion. Perhaps they see themselves stuck between a rock and a hard place—they reject dualism as unscientific; however, if they choose materialism over dualism, perhaps free will must be abandoned. The reasoning seems to go something like this—if we can't see how free will could possibly arise, maybe it just doesn't exist. Others avoid this extreme position and attempt to stake out a middle ground; some in materialistic camps claim that future science will somehow show that materialism and free will are compatible. As discussed in chapter 2, these vexing questions have been around for centuries; at best we can only hope to cast a little light on the mysteries.

Both quantum mechanics and classical chaos have served to toss extreme determinism into history's trash bin, as will be discussed in chapters 9 and 10. Still, the modern physical principles just don't seem to endow us with free will. If even the lofty field of quantum mechanics can't save us, what can? In short, if we really believe in free will, must we reject some of the most "scientific" interpretations of consciousness? After all, what is free will other than the influence of the mind on brain and body?

Having applauded the excellent scientific work by Dehaene in the above sections, I now take issue with some of his expressed philosophical positions. As I've said earlier, one of life's intellectual strategies is to study the work of someone smart and articulate with whom you disagree—as the

Buddhists say, "Your enemy is your best teacher." Here are two examples from Dehaene:[9]

- *Consciousness is just brain-wide information sharing.* Later I will argue that the innocent-looking label "information" raises profound questions. If we define *information* narrowly, this statement seems to me to be just an attempt to rid ourselves of the hard problem on the cheap—the problem is too hard so let's pretend there is no problem. On the other hand, if we define *information* very broadly, let's say "that which distinguishes one thing from another," this statement is just a tautology, an expression that must always be true because of the chosen word definitions.
- *The conception of free will requires no appeal to quantum mechanics and can be implemented in a standard computer.* Dehaene is probably right to say that quantum mechanics doesn't seem to help us much with the question of free will. However, to claim that a computer can be endowed with this mysterious entity called free will, based on no evidence whatsoever, requires an enormous leap of faith. As I see it, we have no real idea of the origin of free will in humans, so how are we supposed to inject this ghostly entity into a computer?

In Libet's experiments concerned with free-will issues, subjects were asked to perform tasks "whenever they felt like it." When they did perform the task, some EEG correlates of their choices apparently occurred before the subjects actually made their choices. In other words, these EEG "go-signals," apparently coming from the unconscious, were recorded before the subjects reported making the conscious decision to act, perhaps implying a lack of free choice. In short, the subjects seem to be indicating something like this—"I didn't really mean to push that button; my unconscious made me do it." In contrast to this interpretation, my colleague Richard Silberstein suggests that it is more likely that the subjects simply set up unconscious "agents" to respond to some random internal event, as indicated in figure 5-3. This picture is based on the continued presence of multiple unconscious systems that interact with the conscious self. The subject's motor (muscle) response can occur through either the conscious (solid arrow) or unconscious pathways. Perhaps such an unconscious neural event originates in the basal ganglia of the midbrain, as this system plays a role in timing, motor responses, and so forth. In this interpretation, no violation of free will need occur.

Several other mechanisms appear consistent with both free will and Libet's studies. First, the conscious will seems to have 100 milliseconds or so to veto any unconscious impulse. Perhaps more important, it seems likely that our unconscious is placed under prior constraints by the conscious mind, depending on the expected external environment. Unconscious agents can be established by the conscious mind well in advance of possible events. If the conscious mind makes a decision to drive a car in traffic, it apparently places prior constraints on impulsive action by the unconscious. The impulse to suddenly rotate the steering wheel 90 degrees while traveling at 70 mph is unlikely to occur. On the other hand, the unconscious impulse to slam on brakes is allowed. In an emergency like a child running in front of the car, the foot can begin to move off the accelerator in about 150 milliseconds, much too fast for the conscious mind to initiate the act. Actual awareness of the child does not occur for another 350 milliseconds, even though there is no conscious awareness of any delay. The free-will process then includes our choices to set up unconscious agents in advance as well as either allowing or prohibiting actions (represented by the dashed arrow in figure 5-3).

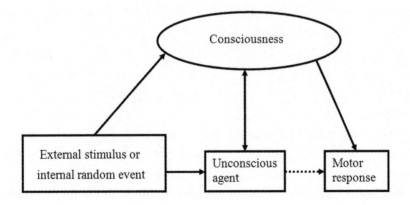

Fig. 5-3.

I am suggesting that the conscious mind is in continuous communication with the unconscious, as represented by the double-ended arrow in figure 5-3, thereby providing updated information on the external environment and constraints on unconscious action. Sometimes we are aware of this communication as an internal dialogue; most of the time this occurs below our level of awareness. Our free will apparently requires ongoing partnerships between conscious and unconscious systems. In special circumstances like playing musical instruments, engaging in sports, or driving a car, we recruit specialized unconscious agents with the ability to carry out certain acts quickly without conscious "permission." In the example of driving a car, an important aspect of free will is the freedom to allow unconscious car braking in an emergency. Our mental health, not to mention freedom from prison, depends on proper control of such impulsive acts. In summary, *free will involves far more than impulsive decisions; free will carries substantial temporal depth.*

5.7 ALZHEIMER'S DISEASE

Dementia, also called *senility,* refers to a broad category of brain diseases that cause long-term reductions in thinking and memory abilities.[10] Dementia diagnoses occur if the observed declines in a person's usual mental functioning are greater than one would expect due to normal aging. The most common type of dementia is Alzheimer's disease, a neurodegenerative disease that usually starts slowly but gets much worse as time passes. No treatments are known that can stop the downward progression; this is a tragic condition for the patient and also for many of the caregivers, often close family members. If your spouse gets Alzheimer's, you are both in trouble.

From the lofty and emotionally detached perspective of science, Alzheimer's provides an interesting example of the different levels of consciousness that humans may experience. The progression of the disease is divided into four stages—a sequence of mental and physical impairments:

- *Pre-dementia.* Mild cognitive difficulties may surface more than five years before a formal diagnosis of Alzheimer's, the most common symptom being short-term memory loss. Former president Ronald Reagan was diagnosed with Alzheimer's in 1994, six years after his term ended. However, some of us are old enough remember a provocative video, shot while he was still president, in which his wife, Nancy, is standing beside him. Apparently, his handlers made the mistake of allowing one of the microphones to come too close. When questions were asked by the press, Nancy can be heard whispering one answer after another to Reagan, who then parrots the answers like a robot. This was a guy with the power to start a nuclear war; remembering that scene still sends chills up my spine.
- *Early-stage.* The early stages of Alzheimer's are difficult to diagnose, but increasing impairment of learning and memory eventually leads to a diagnosis based on shrinking vocabulary and impoverished oral and written language. The person is usually capable of communicating basic ideas adequately but requires help with moderately demanding tasks, such as reading a road map or finding a favorite television show.
- *Moderate.* Progressive deterioration eventually hinders independence, with the person becoming unable to perform common activities of daily living. Complex physical actions become less coordinated, so the risk of falling increases. Memory problems worsen, and the person may fail to recognize close relatives.
- *Advanced.* The person is completely dependent on caregivers. Language is first reduced to simple phrases or single words, eventually leading to complete loss of speech. Death is usually caused by some external factor like pneumonia.

My ninety-three-year-old mother, while still mentally sharp, spent a short time in a healthcare facility because of a bone fracture. The facility also held Alzheimer's patients in various stages. One old guy never spoke, made eye contact, or indicated any of the other usual evidence of awareness. But he did have a "hobby"—he spent most of his waking hours pushing a cart around the hallways. If you were in his way, he would politely go around you, but he would also remember that he had not covered your particular spot. Later he would come back and pass over the area; it seemed like he thought he was mowing a lawn. Apparently, some of his unconscious systems remained in operation, but he lacked anything close to normal consciousness.

Alzheimer's disease is characterized by loss of neurons and synapses in the cerebral cortex and certain subcortical regions. Brain tissue contains excessive abnormal tissue called *plaques* and *tangles* that may only become evident during autopsy. People who engage in intellectual activities like reading, board or computer games, music, or regular social activities show a reduced risk for Alzheimer's. Apparently, some life experiences may result in more-efficient neural functioning, thereby providing the person with a *cognitive reserve* that delays the onset of symptoms.

5.8 SCHIZOPHRENIA

Schizophrenia is a mental disorder characterized by abnormal social behavior and failure to maintain full touch with reality.[11] Symptoms may include confused thinking, false beliefs, and hallucinations. Impaired social engagement and emotional expression may also be evident. Symptoms associated with schizophrenia surface on a continuous scale and must reach a certain severity before a medical diagnosis is confirmed—a decision based on observed behavior and the person's own reported experiences. Individuals may suffer disorganized thinking, hear voices, or experience symptoms of paranoia, perhaps in the form of a delusion that someone is out to "get them" in some way. Involuntary hospitalization is sometimes called for, although antipsychotic medications may allow such hospital visits to be short.

Contrary to much public perception, the label "schizophrenia" does not imply a split personality or multiple personality disorder—a separate condition with which it is often confused. Rather, the term indicates a "splitting of mental functions." In this context, recall our football-fan metaphor, where extreme local states involve only conversations between neighbors. A healthy consciousness is associated with states of moderate or balanced localization. Schizophrenia is often described in terms of *positive* and *negative symptoms*. Positive symptoms include delusions, disordered thoughts and speech, and hallucinations. Negative symptoms are deficits of normal emotional responses or other thought processes, and they are less responsive to medication.

Late adolescence and early adulthood are peak periods for the onset of schizophrenia; in about half the cases it's been associated with subtle differences in brain structures, perhaps white matter deficits, as discussed in the next section. The disease is also thought to be associated with faulty neurotransmitter systems, perhaps preventing the normal functional integration of widespread brain regions required for healthy brain function.

Years ago, I ran into an old friend, "Jane," whom I had not seen for several years. I recalled that her sister was schizophrenic and that genetics put Jane at increased risk. When she saw me, the first thing out of her mouth was, "Oh no, they got you too!" It wasn't just what she said; it was her facial expression and how she said it—a chill went up my spine. I knew right away that Jane had also come down with schizophrenia.

Another disease closely associated with schizophrenia is *Capgras syndrome*, a delusion that causes persons to believe they are living in a world of imposters. An afflicted person sees the people around him, particularly close loved ones, as identical doubles of the people he knows. This condition is like the horror movie syndrome well represented by the movie *Invasion of the Body Snatchers*. The victim of this delusion is aware of his abnormal perceptions, but even when he sees normal behavior around him, he is unable to shake the strange perception that parents, siblings, close friends, and even his spouse seem to be impostors. He may see both the "real" person and their "doubles" at different times; he may even extend the delusion to objects and animals. Capgras syndrome is associated with damage to the posterior right hemisphere of the brain, an area involved with face recognition.

5.9 WHITE MATTER MATTERS TO MENTAL HEALTH

Much of brain science has to do with *structural* and *functional connections*.[12] Structural connections are just what the label implies, consisting mainly of axons that connect one brain location to another. These kinds of connections are "hardwired," although they can change on moderate or long timescales. Functional connections are a much less well-defined; a pair of brain locations is said to be functionally connected if some measured brain activity at one location can predict something about measured activity at another location. Nothing needs to be known about the cause of such correlations when they are established experimentally. The main point is that these functional connections can be turned on and off rather quickly, on timescales consistent with experience and behavior. Whatever their cause, they are closely involved with the brain's dynamic patterns.

Functional connections occur in sharp contrast to structural connections, although the former would seem to be strongly constrained by the latter. As outlined in chapter 3, the white matter layer just below the cortex is composed of myelinated axons; in human brains, most of these axons connect different regions of the cortex to one another. Additional axons connect the two brain hemispheres; still others go back and forth to the thalamus, an egg-shaped structure that sits to the top and side of the brainstem. Myelin consists of special cells that wrap around axons and increase action potential speed, thus the timing of neural events is strongly influenced by myelination. In other words, the timing of signal traffic between distant cortical regions appears to be critical for optimal mental performance and learning.

A central dogma of synaptic plasticity during learning is the importance of *firing coincidence* among multiple synaptic inputs to any neuron. It is generally expected that action potential inputs that arrive at about the same time will form strengthened synaptic connections. In contrast, inputs that arrive non-coincidently will tend to be eliminated. In short, *the neurons that fire together, wire together*. The molecular mechanisms for synaptic plasticity according to coincident firing have been studied extensively; however, the conduction time through axons from pre-synaptic neurons is rarely considered.[13]

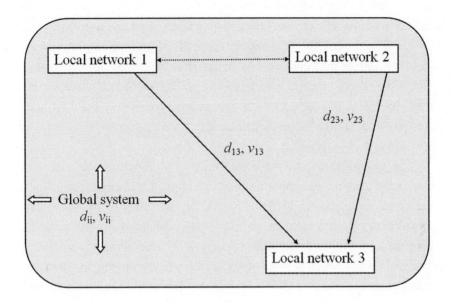

Fig. 5-4.

Figure 5-4 is based on a conceptual framework in which local brain networks are generally embedded in a global system or network. Alternate labels for this system might include "global field" in physics terminology terms or "global workspace" in the parlance of cognitive scientists. Whatever the label, the global system is shown to include state-dependent functional connections between local networks.[14] To demonstrate the general idea of axon-speed flexibility, three "semi-autonomous" local networks are shown embedded in a global system of synaptic and action potentials, somewhat analogous to social networks embedded in a culture. The full figure might represent the entire brain or only part of the cerebral cortex; the local networks might occur at the regional, macrocolumn, or smaller scales.

The label "semi-autonomous" indicates that, to first approximation, each local network is disconnected or only weakly functionally connected to other networks so as to produce truly "local" dynamic behavior. The dashed line between networks 1 and 2 represents a very weak connection. By contrast, strong interactions between local networks will result in new, larger-scale networks with dynamic behaviors that are generally much different than the original smaller-scale behaviors. The inputs from networks 1 and 2 to network 3 will arrive at approximately the same time only if the separation distances (d_{13}, d_{23}) and axon speeds (v_{13}, v_{23}) yield nearly identical time delays between local systems. Thus, synchronous arrival from unequal distances requires modified axon speeds, perhaps by myelin plasticity. By contrast, connections in the global system may have arbitrary strengths and a broad range of distances, d_{ij}, and axon speeds, v_{ij}, between pairs of cortical locations ij.

The relationships between anatomical and functional connectivity involve the spatial and temporal scales at which all measures are obtained. Axon connectivity in white matter axons may be estimated from the injection of chemical tracers transported along cell projections in the living brains of animals. In humans, structural connectivity is accessible by postmortem examination of dissected tissue or noninvasive brain imaging methods like *diffusion tensor imaging* (DTI). In DTI, MRI is used to measure the preferred direction of water diffusion in each brain voxel, thereby providing estimates of major white matter tracts at the 1 mm scale.[15] This approach is based on the idea that the direction of fastest diffusion indicates voxel-fiber orientation, where here the label "fiber" applies to bundles of many parallel axons. While DTI demonstrates impressive technology, it currently falls far short of the resolution required to view most individual axons. Diameter histograms of human cortico-cortical axons are peaked in the 1 μm (micrometer or micron) range; that is, about one thousand times smaller than the 1 mm resolution of DTI. Human white matter actually contains about 10^{10} cortico-cortical axons, far more than the number of tracts revealed by DTI. Thus comprehensive maps of axon connectivity at multiple mesoscopic and macroscopic scales may be years away. Unfortunately, this "inconvenient truth" seems to have been substantially underappreciated in some connectome-related publications.

It's not hard to imagine that disruption of synchronized timing due to white matter deficits may lead to mental and other brain diseases. Myelin controls action potential speed, and the synchrony of impulse traffic between distant cortical regions may be critical for optimal mental performance and learning. A broad range of psychiatric disorders, including schizophrenia, chronic depression, bipolar disorder, obsessive-compulsive disorder, and post-traumatic stress disorder, has recently been associated with white matter defects. Also included are developmental cognitive and emotional disorders including autism, dyslexia, and attention-deficit hyperactivity disorder. The evidence for white matter involvement consists of gene-expression studies, several different kinds of brain imaging studies, and histological analysis of postmortem tissue, as summarized in the publications of R. Douglas Fields.[16]

5.10 SUMMARY

Unconscious mental functions take place faster than conscious mental functions. Conscious awareness requires several hundred milliseconds to develop; our sense of the "now" is distributed over about one half of a second in the past. Consciousness is associated with the total dynamic activity of distributed cortical networks rather than any one network node or by any one instant in time. Experiments designed to study consciousness may employ visual images flashed to subjects. These pictures can be made to vanish from consciousness, or they can be presented such that they are perceived by subjects only half the time. With this kind of experimental design, consciousness becomes a manipulated variable—subjects report seeing the image in some cases but not in others. The essential scientific question is then how dynamic patterns differ between the two cases.

Four reliable signatures of consciousness are outlined: fMRI activity is amplified in some brain regions when the awareness threshold is crossed. Conscious access to visual images also results in a scalp-recorded event-related potential called the P300. Bursts of high-frequency 40 Hz oscillations recorded from inside the brains of epileptic patients may follow generation of the P300. These gamma rhythms are apparently synchronized across some brain regions. Thus, correlated EEG activity, including gamma rhythms as well as slower oscillations, appears to be an important companion to conscious perception.

Electrophysiology refers to studies of electrical properties of tissue over a wide range of spatial scales, ranging from individual cells to scalp-recorded EEG. *Microelectrodes* detect actions of single neurons. *Local field potentials* (LFP) are obtained with larger electrodes, which record the space-averaged activity of multiple neurons. *Electrocorticography* (ECoG) indicates recording from electrodes placed on the cortical surface. *Electroencephalography* (EEG) involves recordings from the scalp surface; its subcategories include *spontaneous EEG* like alpha rhythms and event-related potentials like P300.

In experiments employing implanted microelectrodes in epilepsy patients, some neurons are surprisingly selective when patients are presented with visual images. Neurons may react only to specific pictures, places, or people; they fire only when conscious perception of a specific image occurs. One neuron may fire only in response to a picture of Einstein; another may be a "Bill Clinton neuron" and so forth. These *face-specific neurons*, or *grandmother cells* in classical terminology, do not respond to pictures of other persons.

Some of the differences between "fast" versus "slow" rhythms as consciousness signatures are due to dissimilarity between human and animal EEGs. However, major differences result from recording at different spatial scales. Gamma rhythms are found inside the skull, but gamma rhythms are very difficult or perhaps impossible to measure reliably at the scalp. Slow EEG frequencies like delta, theta, alpha, and beta rhythms are much more prominent and reliable in scalp recordings. This aspect of EEG is explained by analogy with the nested lights of a fanciful Christmas tree. The lights tend to flicker with mixtures of many different frequencies at all scales; however, *the rhythms actually observed depend critically on the spatial scale adopted in the experiment*. Scalp recordings tend to remove highly localized fast frequencies; intracranial recordings tend to downplay widespread slow frequencies.

Free will is our apparent ability to choose between different courses of action. It seems impossible to reconcile the existence of free will with strong or even moderate versions of determinism, in which future events follow directly from the current state of things according to the laws of physics. Some philosophers and scientists claim that free will is an illusion, perhaps reasoning as follows—if we can't see how free will could arise, then maybe it doesn't exist. Others occupy middle grounds, claiming that future science will show that materialism and free will are compatible. While the modern scientific fields of classical chaos and quantum mechanics reject strong determinism, such science has not yet provided a path to solve the deep mystery of free will.

One interpretation of the interactions of free will with unconscious systems is as follows: The conscious mind is in continuous communication with the unconscious, providing updated information on the environment and constraints on unconscious action. Our free will apparently requires partnerships between conscious and unconscious systems. In special circumstances like playing musical instruments, engaging in sports, or driving a car, we recruit specialized unconscious agents with the ability to carry out certain acts quickly without conscious "permission." Our mental health depends on proper conscious control of such impulsive acts; free will has substantial temporal depth.

Relationships between mind and brain are often studied in the context of various brain diseases. *Dementia* refers to a broad category of diseases that cause long-term reductions in mental abilities. The most common type is Alzheimer's disease, which usually starts slowly but gets worse as time passes. No treatments are known that can stop the downward progression; this is a tragic condition both for the patient and also for caregivers.

Schizophrenia is a mental disorder characterized by abnormal social behavior and failure to maintain full touch with reality. Individuals may suffer disorganized thinking, hear voices, or experience symptoms of paranoia. Contrary to public perception, the label "schizophrenia" does not imply a split personality—a different condition with which it is often confused. Rather, the term indicates a splitting of mental functions.

Much of brain science has to do with structural and functional connections. Structural connections consist mainly of axons that connect one brain location to another. These kinds of connections are hardwired, although they can change on moderate or long timescales. The proper timing of internal signals is essential to healthy brain function. Learning is believed to occur when multiple action potentials arrive nearly simultaneously at a postsynaptic neuron, thereby strengthening the connection. Myelin controls action potential speed and the synchrony of impulse traffic between cortical regions. A broad range of psychiatric and developmental disorders involves myelin defects in white matter. In the next chapter we will see that brain diseases are also closely associated with abnormal brain rhythms, caused by some combination of neurotransmitter, myelin, and other functional and structural defects. I will argue that brains require proper "tuning" in order for healthy consciousness to emerge. Such tuning is closely analogous to the usual tuning of musical instruments required to produce the beautiful music of "healthy" orchestras.

Chapter 6

RHYTHMS OF THE BRAIN

6.1 THE MUSIC OF CONSCIOUSNESS

This chapter's title is more than just a catchy tag; brain rhythms have long provided highly reliable signatures of consciousness.[1] Such scientific studies typically involve the brain's many and varied rhythms, often recorded from the scalp with EEG. As suggested symbolically by figure 6-1, the brain's electric (EEG) and magnetic (MEG) fields both extend a little outside the head, but these external fields fall off rapidly with distance from the scalp.[2] For our purposes, the labels "rhythm" and "oscillation" are used almost interchangeably. The word *rhythm* is typically associated with music, while the tag *oscillation* often applies to physical systems. But in point of fact, music is produced by physical systems, usually by mechanical vibrators like pianos, guitars, and vocal cords. *Vibrators* are found in both man-made and natural systems; they appear just about everywhere. By linking the labels "rhythm" and "oscillation" we imply that they are nearly the same thing; this tight link is appropriate to facilitate our use of musical analogues and metaphors.

The reasons for our emphasis on rhythms in this chapter are many and varied. Since first recorded in the 1920s, EEG rhythms have been closely related to a broad range of mental functions, including information transfer, perception, muscle control, memory, and general conscious state. If your brain doesn't produce brain rhythms, you are one dead duck—one of the medical tests for *brain death* is a "flat EEG," that is, the absence of electrical rhythms on the scalp. Electrical oscillations are found at many locations in the brain and at different levels of tissue organization, as outlined in chapter 5. The cell assemblies creating these rhythms seem to range from single neurons to nested neural networks of different sizes, and apparently even to globally generated phenomena.

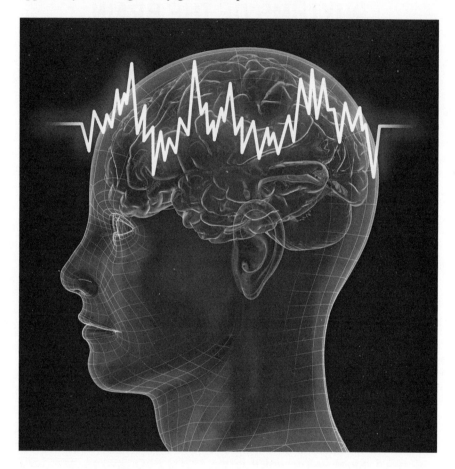

Fig. 6-1. © Can Stock Photo Inc. / beawolf.

Our discussions of brain rhythms and their close connections to mental processes make extensive use of analogue and metaphorical systems. These might be electrical, mechanical, molecular, social, and more. Remember that here we employ the label "metaphor" to convey the meaning of some concept in a manner that is more easily visualized than the original concept, but the metaphor need bear no genuine relationship to the original concept. In contrast, an "analogue" possesses many of the same properties enjoyed by the more-complex thing that it represents, but in a simplified manner that's easier for us to grasp. In short, our use of the label "analogue" implies a stronger correspondence between systems than does "metaphor."

Chapter 3 employed the football-fan metaphor to demonstrate distinctions between brain states of functional localization and states of global coherence. It was argued that healthy brains tend to operate between these extreme conditions. An alternate metaphor was chosen in chapter 5—the flickering lights in our fanciful magic Christmas tree demonstrated that distinct dynamic patterns of brain activity are measured at different spatial scales. Still another brain metaphor is employed in this chapter—the *orchestra*, a system consisting of the vibrating mechanical structures better known in lofty intellectual circles as *musical instruments*. We will, in fact, argue that this musical "metaphor" may often rise to the level of "analogue."

The label "vibration" refers to oscillations of air pressure or elastic solids like the guitar string. The vibrating object then acts as the source of sound that moves through the air or some other medium that supports wave propagation. The vibrating object might be a person's vocal cords, the string or the body of a guitar, or the diaphragm of a computer speaker. Many vibrating objects can create sounds, even the proverbial tree falling in a forest. The sound might be musical or noisy; but regardless of the quality, sound waves are created by vibrating objects. By contrast, many systems produce oscillations of other things that have little to do with mechanical vibrations. In brains, oscillations of synaptic activity in large tissue masses are recorded on the cortex or scalp as oscillations of electric fields. That said, the general properties of rhythms apply to a wide range of systems, and these properties are more easily visualized in mechanical systems like musical instruments—we will make good use of this correspondence.

Our *rhythmic brains* are, in several ways, much like magical orchestras, ensembles of numerous musical instruments.[3] A typical orchestra consists of four different sections containing string, brass, wind, and percussion instruments. These four sections may contain violins, horns, flutes, and drums (a brass section is depicted symbolically in figure 6-2). Our metaphorical orchestra has no conductor and no conscious musicians;

the fanciful instruments play by themselves and produce a wide range of rhythms. Flutes tend to vibrate the air at a single frequency, producing a pure tone. By contrast, tubas vibrate at a special mix of frequencies that produce complex tones. Orchestra rhythms are generated at several levels of organization. For example, small-scale oscillations originate with the strings of the violin; the strings then stimulate a different mix of larger-scale rhythms produced by the violin body, these rhythms are called the violin body's *overtones*.

Fig. 6-2. © Can Stock Photo Inc. / Petrafler.

An even-larger-scale mix of rhythms may be generated by a group of instruments forming each orchestra section. To make this analogy consistent with genuine brain rhythms, we may imagine that each musical section produces novel frequencies in addition to the mix generated by individual instruments. That is, new frequencies emerge in each magical-orchestra section. This could be accomplished mechanically in a real string section by physically connecting several of the violin bodies with some sort of fanciful "bridge." The resulting "deep-throated violin" would then be expected to produce rhythms of very low frequency since larger stringed instruments tend to produce a lower range of frequencies. Similar arguments might be applied to the brass, wind, and percussion sections. Our perception of the quality of the sounds produced by the orchestra is determined by the mix of frequencies produced. By analogy, the mix of rhythms produced in the brain is closely related to our conscious and unconscious states. *In order to be mentally healthy, we must be properly "tuned."* Does this description sound like some flaky mumbo jumbo dreamt up by meth addicts? Maybe it sounds that way, but in the end, reality is revealed by genuine experimental science, not preconceived notions of pseudoscience. Stay tuned.

Our sensation of frequency is referred to as the *pitch* of a sound. Each pitch corresponds to a particular frequency, expressed in vibrations per second or Hertz (Hz). The human ear is capable of detecting sound waves with frequencies ranging between approximately 20 Hz to 20,000 Hz. Dogs can hear sounds up to 45,000 Hz, and cats are even "faster" little beasts, hearing sounds up to 85,000 Hz. Any two sounds whose frequencies occur in a 2:1 ratio normally result in pleasant sensations when heard by humans. That is, two sound waves sound good to most of us when played together if one sound has twice the frequency of the other; the reasons why we like this simple frequency ratio are unknown.[4] Similarly, two sounds with a frequency ratio of 5:4 also sound good when played together. Why? Nobody knows. Music appreciation is, of course, a very personal aspect of our consciousness, reflecting a broad range of emotions and opinions. Mark Twain once offered the following observation about a famous composer—"Wagner's music is better than it sounds."[5] But apparently not all agree—Wagner's music is actually banned in Israel because of his anti-Semitism. Note, however, that Wagner died fifty years before Hitler's rise to power—so this issue raises interesting free-speech issues that are beyond the scope of our discussions.

The frequencies at which objects like musical instruments tend to vibrate when disturbed by external forces are known as the *natural frequencies* of the object. For example, the violin shown in figure 6-3 has four strings (G, D, A, E), and each string has its own set of natural frequencies, consisting of the *fundamental* (lowest) frequency and its *harmonics*, called *partials* by musicians. The strings, however, produce very little sound; most of the sound is produced by vibrations of the violin body. The violin body responds selectively at its own natural frequencies, its fundamental and its overtones. In short, the violin strings generate special frequencies that drive the body, but the body itself responds selectively, demonstrating larger-scale *resonance*.

The selectivity of the strings and body indicate the phenomenon of resonance, the tendency of systems to respond strongly only to certain frequencies. The violin body responds to oscillations of all four strings, but it responds most strongly to oscillations of the A string, which has a fundamental frequency of 440 Hz. Each string's set of natural frequencies, the fundamental and higher harmonics, may be adjusted by turning knobs at the end of the violin, thereby changing the tension in selected strings. Tighter strings produce higher fundamental and harmonic frequencies.[6]

Our metaphorical ghost orchestra consists of musical instruments but no conductor or other conscious musicians. Any air or ground disturbance might stimulate the instruments so that they produce vibrations at their natural frequencies. The resulting sound is not likely to be pleasant to some imagined godlike listener, but this is of no concern to us. These basic ideas about musical sounds, natural frequencies, resonance, and so forth apply to most oscillating systems, including the electromagnetic fields picked up by your cell phone, light sensed by cells in your retina, atomic events, and neural networks of various sizes and brain locations.

In the orchestra, rhythms generated by the different musical sections are produced at several levels of organization (spatial scales). In the case of the stringed instruments, there are at least three such levels, the strings, the instrument body, and the string section containing multiple instruments. In contrast to string harmonics, the higher natural frequencies of the violin body are not harmonics; that is, they are not multiples of the fundamental. Thus, new notes may be generated by the violin body, which fail to match any of the string's natural frequencies; the higher natural frequencies of the body are known as its *overtones*.

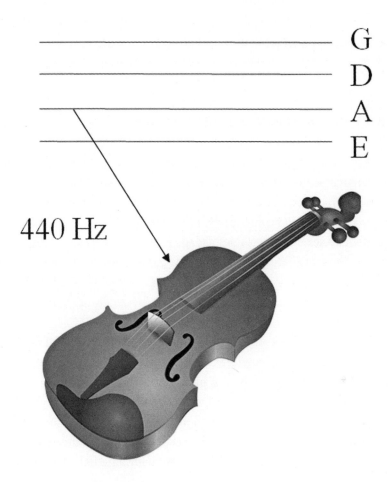

Fig. 6-3. © Can Stock Photo Inc. / Linda B.

Similarly, brain rhythms are generated at several or perhaps many organizational levels, ranging from the single neuron to the entire cerebral cortex and perhaps even the entire brain. The violin's frequencies are usually controlled by turning knobs to change string tension, but they might also be controlled mechanically at the level of the violin body. This could be accomplished with any sort of clamp that interferes with the violin body's standing wave patterns, which are responsible for its resonance behavior. That is, the fundamental and overtones are caused by the vibrations of the corresponding standing wave patterns in the violin body. In comparison, brain frequencies are controlled by various chemical neuromodulators that interact with neural networks at various scales, perhaps including large-scale *standing wave patterns*, as discussed in chapter 7.

6.2 RHYTHMS RECORDED ON THE SCALP

We continue our allegory of "brain music" with the actual rhythms recorded from the human scalp, called the *electroencephalogram* or *EEG*. These oscillations were first discovered in the 1920s and are the most widely studied of all human rhythms. EEG consists only of the global brain rhythms and large-to-intermediate-scale rhythms that manage to reach as far as the scalp—the reasons for this selectivity are closely associated with the synchrony of some *brain sources* (generators of electric fields). In contrast, oscillations generated at smaller scales can only be studied with electrodes implanted inside the skull, either on the brain surface or inside the brain itself.

Imagine yourself as a prominent EEG scientist. As part of your research project, you recruit "Sue," a willing experimental subject represented here by the mannequin in figure 6-4. You place a special electrode cap on Sue's head, a simple and painless procedure. The experiment might go just a little easier if Sue were bald, but you don't even think about asking her to shave her head. Sue is dedicated to science, but not that dedicated. Not to worry, modern EEG technology deals nicely with long hair. The cap may hold a hundred or more embedded electrodes that fit snugly against Sue's scalp; the electrodes sense electric potentials and scalp currents generated inside her brain. Sue's EEG is recorded overnight while she lies in a comfortable bed. EEG is transmitted to an isolated location with a computer screen showing her ongoing brain rhythms. In addition, a video camera sends images of Sue's face to the computer monitor. Various physiological measures like muscle activity and eye movements are also displayed.

Even if you are unfamiliar with EEG, its general relationship to Sue's mental state will soon become apparent. During *deep sleep* her EEG has larger amplitudes and contains much more low-frequency content, only about one to four oscillations per second, called *delta rhythms*. During rapid eye movement sleep (REM), when her closed eyelids reveal underlying eye movements associated with dreaming, Sue's EEG contains much more fast-frequency content. When Sue is awake and relaxed with closed eyes, oscillations repeating about ten times per second are evident; these are known as *alpha rhythms*. More-sophisticated monitoring allows for identification of distinct *sleep stages, depth of anesthesia, seizures,* and other *neurological disorders*. Similar scientific experiments reveal EEG correlations with cognitive tasks like *mental calculations, working memory,* and *selective attention*. EEG information is, however, mostly limited to general mental states. We will certainly know whether or not Sue is conscious, but the EEG will not tell us what Sue is thinking or if she is dreaming about her boyfriend or perhaps a vacation in the Caribbean.

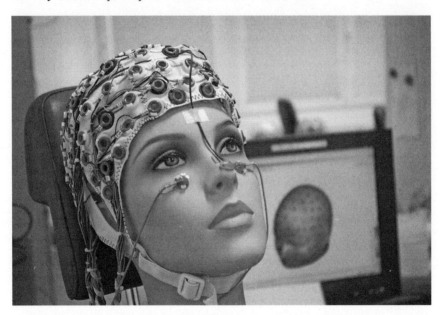

Fig. 6-4. © Can Stock Photo Inc. / neurobite.

Many scientists are now so accustomed to these EEG correlations with brain state that they may forget just how remarkable they are. EEGs provide very large-scale and reliable measures of *brain dynamic behavior*. By this label we mean *dynamic patterns*, the spatial patterns of electric fields and currents over the scalp that change rapidly with time. The general idea of dynamic behavior in complex systems is demonstrated nicely by a series of weather maps like the snapshot shown in figure 6-5. This kind of changing pattern is analogous to dynamic patterns of electrical activity on the scalp or brain surface. For example, the epilepsy diseases are sometimes labeled "storms in the brain," analogous to the hurricane over Florida shown in the figure.

Fig. 6-5. © Can Stock Photo Inc. / CarolinaSmith.

A single scalp electrode provides estimates of neural activity averaged over tissue masses containing roughly between ten million and one billion neurons. The space averaging of brain potentials in scalp recordings is forced by current spreading through the head. In a similar manner, the weather map of figure 6-5 represents space averages over many moderate-sized air masses located in different regions. Much more detailed information about the brain may be obtained from intracranial recordings in animals and epileptic patients—such electrical activity may be recorded from the brain surface as electrocorticograms (ECoG) or within its depths as a local field potentials (LFP). But intracranial electrodes implanted in living brains provide only very sparse spatial coverage, thereby failing to record the "big picture" of brain function. Depending on spatial scale or the size of neural networks under study, intracranial recordings can see the proverbial trees and bushes, and maybe even some of the ants on branches. By contrast, EEG sees much of the forest, but misses most of the detail. As in all experimental science, measurement scale is a critical issue that must be considered carefully when interpreting data. Studying this book with a microscope may cause you to miss the large-scale information content of its words. In contrast to the apparent views of some, *intracranial data generally provide different information, not more information, than is obtained from the scalp.*

6.3 RHYTHMIC THINKING

Before we dig more deeply into relationships between brain rhythms and consciousness, let's make sure that we understand some of the basic ideas that apply to all kinds of rhythms or oscillations, whether they involve brains, sunspots, weather, earthquakes, music, traffic noise, light, cell phones, or any of the many kinds of oscillations that influence our daily lives. Don't worry; this discussion will not involve much in the way of technical detail. I just want to ensure that we all have a good grounding in Rhythms 101 before going forward.

Each of the six sine waves (*sinusoidal oscillations*) shown in figure 6-6 consists of five oscillations in the one-half-second periods shown, or ten oscillations per second. Each of these waveforms has amplitude equal to one (in arbitrary units). The unit of oscillation frequency is given the name Hertz; thus, we call each of these rhythms "10 Hz oscillations." The two waveforms in the uppermost plot differ only by a small *phase difference*; that is, the peaks of the dashed curve occur a little earlier than peaks of the solid curve. In contrast, the lower two waveforms are more *out of phase*; the dashed waveform in the lowest plot is fully out of phase with the waveform given by solid line. Every so-called sine wave is fully determined by just three features—its amplitude, frequency, and phase. When the phases line up, or nearly so, as in the top figure, we say that the two signals are *synchronous*. Some neuroscientists make the unfortunate error of equating *synchrony* with *coherence*. These are closely related concepts, but they also involve very important scientific differences, as we shall see in chapter 7.

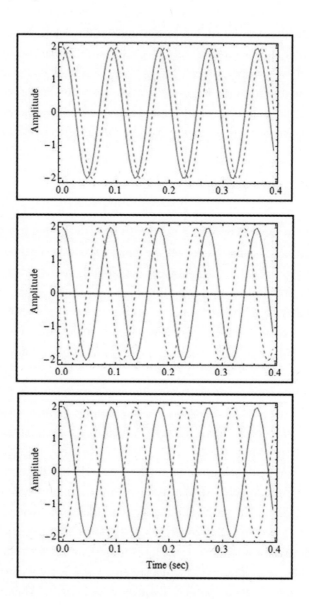

Fig. 6-6.

In many fields, including music and brain science, we wish to know the relative contributions of different frequencies to our signals. Such measures are especially important in EEG work because different *frequency bands* provide selective information about different brain states and specific mental activities. To achieve this goal of picking out individual frequencies from a complex signal, computer algorithms are employed that can decompose any waveform, no matter how complicated, into its composite frequency components. This procedure, called *spectral analysis*, is analogous to atmospheric decomposition of white light, containing a broad range of light colors and frequencies, into distinct rainbow colors, each belonging to a specific frequency band of electromagnetic-field oscillations.

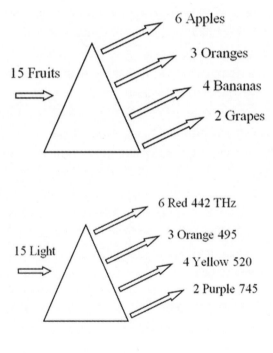

Fig. 6-7.

This general idea is demonstrated in figure 6-7, where we imagine a grocery bag containing fifteen mixed pieces of fruit. The upper triangle might represent some sort of conveyer belt with holes of different shapes used to sort the fruit collection into its component fruits. Similarly, the prism shown in the lower part of figure 6-7 sorts the incoming light into four color components—red, orange, yellow, and purple. This light input to the prism is called "colored light" rather than "white light" because the distribution of color intensity is not uniform, as indicated by the numbers in front of each labeled color. Thus, the particular colored light depicted here contains three times more red than purple, and so on. The light frequencies listed here, corresponding to the center of each color band, are expressed in terahertz (THz), where one THz equals one trillion (10^{12}) Hz. The narrow region of electromagnetic radiation that we humans call "light" consists of very fast oscillations indeed. Special cells in the retina, the *rods* and *cones*, are finely tuned to these superfast frequencies, acting like tiny antennae that allow us to see the world around us.

Our perception of light begins as a quantum *resonance phenomenon*; that is, special cells in our retinas are selectively sensitive to a very narrow range of the electromagnetic fields that bombard our eyes on a constant basis. Figure 6-8 shows the relationship of the narrow frequency band called "light" to other parts of the electromagnetic spectrum. Not shown are the ultraviolet and infrared bands, located just to the left and right, respectively, of the very narrow visible light band. The wiggly line at the top indicates symbolically that oscillations get slower as we move from left to right, but the actual frequency range depicted spans more than twenty factors of ten (10^{20}). Other species are selectively sensitive to somewhat-different light frequencies, but mostly in the same general range, mainly because they evolved in the same atmosphere, which selectively filters the sun's electromagnetic fields.

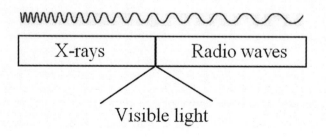

Fig. 6-8.

The EEG community employs its own labels to characterize wave-forms according to frequency band: *delta* (1–4 Hz), *theta* (4–8 Hz), *alpha* (8–13 Hz), *beta* (13–20 Hz), and *gamma* (usually greater than 30 Hz). These categories should be taken with a pound of salt because EEG is typically composed of mixtures of many frequencies. For example, so-called beta activity may actually be composed of beta oscillations superimposed

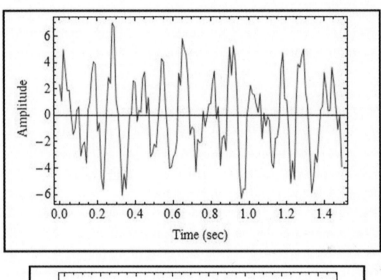

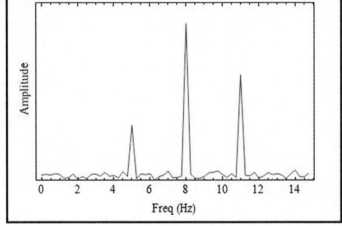

Fig. 6-9.

on larger-amplitude alpha. This confusion originates from times before the 1970s when spectral analysis came into more common use in EEG. Here we demonstrate spectral analysis using the artificial signal shown in the upper part of figure 6-9. The sample signal may seem a bit complicated at first; however, spectral analysis of much a longer time interval shows that it actually consists of just three frequency components, 5, 8, and 11 Hz, plus a little noise. The corresponding relative amplitudes of the frequency components are 1, 3, and 2, as indicated by the heights of the three peaks shown in the lower plot. These amplitudes are analogous to the numbers of specific fruits in our metaphorical fruit basket in figure 6-7.

6.4 NORMAL AND ABNORMAL RHYTHMS

The so-called alpha rhythm is an important human EEG category that actually embodies several distinct kinds of alpha rhythms, usually identified as oscillations at frequencies near 10 Hz. A four-second epoch, consisting of multiple alpha rhythms, in an awake and relaxed human subject is illustrated by the plot over time in the upper part of figure 6-10; the amplitude axis is labeled in microvolts (one millionth of a volt). This tiny signal is about one hundred times smaller than the electric potential measured over the heart (EKG). In order to improve accuracy, the corresponding frequency spectrum in the lower plot is based on a full five minutes of data rather than the four seconds shown in the upper plot. More-sophisticated analyses suggest that this rhythm is probably composed of at least three distinct processes, two intermediate-scale networks plus a global (entire cortex) contribution. The double peak in the alpha band partly reflects this multicomponent interpretation. Other tests, involving distinctive spatial distributions over the scalp and selective reactivity to mental tasks of different alpha frequencies, support this multiprocess interpretation of alpha rhythm phenomena.

Alpha-rhythm amplitudes are typically larger near the back of the brain (the *occipital cortex*) and smaller over frontal regions, depending partly on the subject's state of relaxation. Often, the more relaxed the

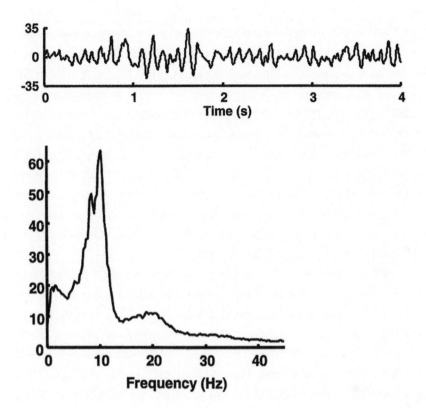

Fig. 6-10. Based on the work of Brett M. Wingeier, "A High-Resolution
Study of Large-Scale Dynamic Properties of Human EEG"
(PhD diss., Tulane University, April 2004).

subject, the larger the frontal alpha. In addition to alpha rhythms, a wide
variety of human EEG activity has been recorded, a proverbial zoo of
"dynamic signatures," each waveform dependent in its own way on time
and scalp location. Normal resting alpha rhythms may have substantially reduced amplitudes with eye opening, drowsiness, or mental tasks.
Hyperventilation and drugs like alcohol may cause a lowering of alpha
frequencies and increased amplitudes. The barbiturates are associated
with increased amplitude of low-amplitude beta activity perhaps superimposed on alpha rhythms. We will revisit alpha rhythms in later sections
of this chapter.

Loss of normal consciousness occurs during sleep, especially the deep sleep more evident early in the night. Sleep and waking are controlled by neuromodulators, special chemicals released by cells deep in the brain. These neuromodulators influence the excitability of target cells in other parts of the brain, thereby acting to *tune the brain* in a manner analogous to tuning the musical instruments in our magic orchestra. As a result of this control by lower brain structures, brain and mind cycle through various stages of sleep every night. The sleeping brain is capable of generating an imaginary inner world, the dream world, which is mostly disconnected from the real world, as discussed in chapter 4. As previously mentioned, the body alternates between non-REM and REM (rapid eye movement) stages; and this latter stage also involves near paralysis of muscles, thereby preventing sleepwalking in most people.

Two famous early EEG pioneers are neuroscientist Herbert Jasper and his colleague neurosurgeon Wilder Penfield, known for his electrical stimulation of cortical tissue, a procedure that sometimes evoked past memories in his epileptic patients. Cortical stimulation of conscious patients is needed even today to identify critical *network hubs* like speech centers to be spared from the surgeon's knife. The brain has no pain receptors, so this procedure is carried out under a local anesthetic. Numerous EEG (scalp) and ECoG (cortical surface) recordings of epilepsy patients were obtained by Jasper and Penfield in the1940s and 50s. This work is summarized in their classic work, *Epilepsy and the Functional Anatomy of the Human Brain.*[7]

Several idealized representations of scalp-recorded normal rhythms are shown in figure 6-11; the vertical scale is 100 mV (microvolts). The resting alpha rhythms occur mainly with closed eyes and have typical amplitudes of about 40 mV, whereas deep sleep and coma may exceed 100 mV. The EEG-like traces shown in figure 6-11 provide consciousness signatures or "fingerprints" of various sleeping and waking states, including different kinds of mental activity. However, genuine EEG typically consists of mixtures of different frequency components. The upper plot might represent deep sleep (stage 3), coma, or anesthesia. The second plot represents theta rhythms that increase in amplitude with certain kinds of mental tasks. The third plot indicates a composite of local and global alpha rhythms, typically recorded in relaxed subjects with closed eyes. Different alpha rhythms with similar or equal frequencies are selectively modulated by different mental tasks. In the lower (beta rhythm) plot, the subject might also have been engaged in some difficult mental task. Gamma rhythms (40 Hz range) on the scalp are very low-amplitude and difficult or perhaps impossible to distinguish from muscle artifact. REM sleep is not shown, but it somewhat resembles the beta trace. Stage 2 sleep consists of mixtures of relatively fast (theta and alpha) and very slow (delta) frequencies, whereas stage 3 (deep sleep) is dominated by delta frequencies near 1 Hz, as shown in the uppermost plot. A normal sequence of the sleep stages as they occur overnight was shown earlier in figure 4-3, indicating that the REM stage associated with dreaming is sometimes viewed as a partial awaking. The brain typically cycles through four or five sleeping cycles each night.

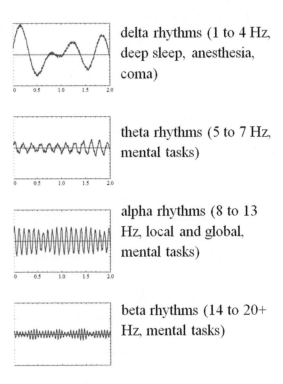

delta rhythms (1 to 4 Hz, deep sleep, anesthesia, coma)

theta rhythms (5 to 7 Hz, mental tasks)

alpha rhythms (8 to 13 Hz, local and global, mental tasks)

beta rhythms (14 to 20+ Hz, mental tasks)

Fig. 6-11.

Progressively deeper states of unconsciousness generally occur with lower-frequency content and larger amplitudes.[8] Note that this general "rule of brain" runs directly counter to the simple Christmas-lights brain where deeper thinking is imagined to occur with more or brighter lights. In fact, the scientific truth is just the opposite—the larger the EEG amplitude and the lower the frequency, the more likely that the subject is unconscious because of deep sleep, anesthesia, coma, or epileptic seizure. The reasons for this outcome have partly to do with the local-global balance described in connection with our football-fan metaphor—a healthy awake brain must operate between the extremes of global coherence and functional localization. Furthermore, the patterns associated with consciousness apparently require the presence of large numbers of inactive neurons. *High dynamic complexity and conscious states seem to require substantial regions of quiet neurons*, as discussed in chapter 9.

The general rule that lower EEG frequencies tend to occur with larger amplitudes has held up quite well in modern studies, although partial exceptions occur. Two such cases are *alpha coma* and *halothane anesthesia*; in both of these cases, large-amplitude alpha rhythms occur over the entire scalp. Thus, the aspect of the "rule of brain" that relates large amplitudes to unconscious states holds; however, these alpha rhythms violate the usual low-frequency connection with unconscious states. This observation serves as still another reminder that the so-called alpha rhythm typically consists of multiple processes along the local-to-global range of alpha-related phenomena.

Another interesting property of the halothane drug is its ability to tune the brain to any dominant frequency between about 16 and 4 Hz by increasing blood concentration of the drug, controlling depth of anesthesia. Deeper anesthesia again results in larger amplitudes and lower frequencies, consistent with our "rule of brain." As we will discuss in chapter 7, halothane rhythms are good candidates for global phenomena like standing waves, with minimal distortion from local networks. Other anesthetics have their own characteristic EEG signatures, typically similar to the deep sleep or coma traces.

Cortical recordings indicate that differences in ECoG dynamic behavior between cortical areas tends to disappear during anesthesia.[9] A large variety of EEG behaviors may be observed depending on depth and type of anesthesia or coma. These include sinusoidal oscillations and complex waveforms (combinations of frequencies) in the delta, theta, alpha, and beta bands. Again, lower-frequency oscillations tend to occur with larger amplitudes in a wide range of brain states. Furthermore, local differences in scalp waveforms tend to disappear with deeper anesthesia, indicating transitions from more local (functionally segregated) to more global dynamic behavior. This observation fits nicely within our conceptual framework of local cognitive networks embedded in global fields or networks, which tend to be more evident when mental processing is minimal. Under "metaphorical anesthesia" our football-fans are all cheering together mindlessly; outside observers can discern little or no signs of intelligence. Mob rule and extreme nationalism provide additional examples. As a citizen of Prague recently characterized the Nazi period to me, "Bands were playing, feet were marching, brains were paralyzed."

The EEG-like traces shown in figure 6-11 indicate only a small part of a much larger "zoo" of EEG rhythms, some having no known clinical correlates.[10] Many abnormal rhythms have been recorded from the cortex and scalps of epileptic patients. These include unusually large rhythms in the delta, theta, alpha, beta, and gamma bands. A 14 Hz rhythm normally occurs during REM sleep; a regular 3 Hz rhythm might occur in epilepsy or in certain comas. A classical epilepsy signature is the "3 Hz spike and wave," which often occurs over the entire scalp. Global spike-and-wave traces typically occur during nonconvulsive absence seizures in which awareness is lost for seconds or minutes. Why do these distinct dynamic patterns occur? The answers are mostly unknown. However, we see clearly that many nuanced differences in conscious states are revealed by very specific and detailed dynamic patterns of EEG, even though this large-scale scalp measure suffers from the limitations of poor spatial resolution.

6.5 SYNCHRONOUS GAMMA RHYTHMS
RECORDED FROM INSIDE THE SKULL

For obvious ethical reasons, human recordings from the cortical surface (ECoG) or from inside the brain (LEP or single-cell recordings) are limited to patients who have electrodes implanted for medical reasons. In addition, intracranial recordings in living animals have long been undertaken to investigate a broad range of scientific issues. Even a limited summary of this work would require an entire book of its own; no such ambitious effort is attempted here. The main point I want to emphasize is that many so-called signatures of consciousness have been discovered with electrophysiology over the past century. As expected in any genuine complex system, the results are often scale-specific. For example, the dynamic patterns of neuron firing rates in local tissues are likely to look quite different than the larger-scale measures of identical mental activity. We suspect that the different patterns are related in some way, but how?

With this caveat in mind, some prominent animal studies of synchronous rhythms are outlined here; similar studies have since been carried out in humans. In the early 1990s, neuroscientist Wolf Singer advanced a provocative idea—different features of an animal's world are represented by distinct cell assemblies (networks) that can be "bound together," that is, functionally connected, by synchronous rhythms.[11] Shortly after, the catchy label "bound by synchrony" entered the neuroscience lexicon. The technical details of such studies need not concern us, but the general idea is this—suppose a monkey is trained to search through a pile of boxes to find a hidden banana when the project scientist rings a bell. This task requires the monkey to integrate functions associated with memory, planning, vision, and directed hand movements.

In these kinds of studies, simultaneous recordings from different areas of animal cortex reveal that cortical areas involved in combined visual, motor, and somatosensory tasks may tend to synchronize local field potentials (LFPs) when the animals prepare for the task and focus their attention. These findings emphasize oscillations at frequencies in the gamma band (typically above 30 Hz). In perceptual-motor tasks (planning and moving), gamma-band synchrony is maintained until the task is completed, at which time the synchronized patterns collapse and give way to lower-frequency desynchronized oscillations. For many neuroscientists, this "binding by synchrony" idea qualifies as a genuine "paradigm shift" by suggesting that synchronized oscillatory activity across different cell assemblies is a critical feature underlying cognition rather than just fixed anatomical connections. In other words, remote neural networks can be made to work together if their rhythms "match" in some way. The analogy to coordinated musical instruments in an orchestra is compelling in this context; the brain tissue that plays together stays healthy together.

In the broader context of the multiscale electrical measures of EEG, ECoG, LFP, and single neurons, the experiments support a conceptual framework where healthy mental activity depends critically on phenomena in multiple, but selective, frequency bands. In humans, such select frequencies can apparently cover a broad range, perhaps something like 1–100 Hz, although specific experimental designs, especially involving spatial-measurement scale, are likely to reveal only (perhaps very small) parts of these spectral signatures. In chapter 7, we will see that a large number of synchrony-related signatures of mental states have been discovered, but these are nearly all in the low-frequency range, that is, less than about 15 Hz. The apparent disconnect between these low-frequency results and the 40 Hz (gamma) signatures outlined above results mostly from the disparate spatial scales involved in different experimental designs. Our takeaway message is this—signatures of consciousness occur at a number of spatial scales and in multiple (but select) frequency bands. In more-colorful language, properly tuned brains seem to orchestrate the beautiful music of consciousness, but brain music is not limited to the popular gamma "tunes." Delta, theta, alpha, and beta rhythms are also on the sentient music charts.

6.6 LOCAL-GLOBAL NERVOUS SYSTEMS

Many old-school psychologists seem to have viewed the brain as a passive, stimulus-driven organ that does not create meaning by itself, but rather reacts to sensory inputs. Sensory processing was imagined to consist mainly of the extraction and processing of features obtained from the environment. One obvious problem with this kind of approach is that it ignores the critical influence of memory. In contrast, modern views consider conscious perception to be a more active and selective process in which stimulus processing is controlled by global brain influences that include memory. These influences shape the dynamic patterns of networks and create *predictions* about future sensory events on an ongoing basis. To cite a matching example, scientists produce *models* of the world that are checked against experimental evidence. Similarly, evolution has produced brains that continuously create models; in healthy brains at least, such models are checked through experience.[12]

In an evolutionary context, the main task of cognition is to guide action, so the test of mental success consists of creating actions adapted to the external environment. Mental activity does not create rigid models of the world; it is subject to the ever-changing environment; thus the resulting constraints must be flexible. This general picture seems to suggest that mental functions require the collective behavior of large neural populations, which are dynamically bound across subsystems. In simple terms, healthy brains operate as complex global systems with embedded local networks, an idea analogous to social networks embedded in a culture. Intelligent behavior presupposes that a mental system can partly detach itself from current stimuli and select only those inputs concerned with the free control of action. In the real world, such processing must be both fast and reliable, allowing predictions about new sensory input and constantly matching expectations against signals from the environment. Memory is clearly a critical feature of this process. In short, the brain both creates and erases models of reality on an ongoing basis. Suppose you walk in a dense forest at night; you may soon become aware of many distinct sounds. Some sounds will be familiar, some will not. Most will seem benign, but some may raise mental warning flags, perhaps even leading to a flight-or-fight response. Each of these sounds contributes to your updated mental model of the environment. In ancient times, one's life depended on continually updating accurate models of the surrounding world. Even in modern times, bad models or models that are out of date often produce adverse consequences.

A model brain seemingly consistent with the top-down and bottom-up dynamic processes implied above is indicated symbolically in figure 6-12. The big gray circle represents a large part of the cerebral cortex or maybe even the entire brain. The many small symbols represent three distinct large-scale networks; that is, identical symbols indicate brain tissue that remains functionally connected (in some way) for at least several hundred milliseconds. Thus, the black triangles represent a network distributed over the back of the brain in both hemispheres. The gray trapezoids are parts of a more frontal network. Each individual symbol suggests local tissue at perhaps the millimeter or centimeter scale; symbols in common indicate a global or regional network, perhaps part of the so-called *global workspace*, in the language of cognitive science.

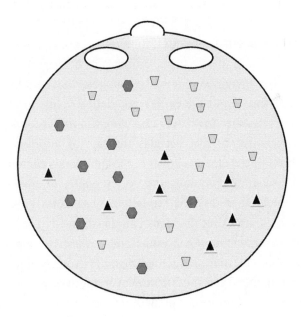

Fig. 6-12.

We have emphasized that consciousness is distributed over time; figure 6-12 suggests that it is also distributed over space within the brain. The spatially distributed aspect of consciousness may be illustrated by a metaphor taken from one of my favorite science fiction stories. In 1957, British astronomer Fred Hoyle published *The Black Cloud*,[13] a provocative little novel read by many of the physics and astronomy students of my generation. Even though this story is more than a half century old, the scientific ideas remain quite up to date, especially in the context of multiple unconscious and pre-conscious systems underlying a large-scale consciousness.

In Hoyle's novel, an enormous black cloud of matter approaches our solar system. The heroic scientists make repeated attempts to study its physical properties by probing the cloud with electromagnetic signals of different frequencies, resulting in cloud *resonant behavior* in apparent violation of known physical laws. After eliminating all other possibilities and following the famous advice of Sherlock Holmes, the scientists are forced to an astounding conclusion—the cloud is conscious! The scientist's next question is whether the cloud contains a single consciousness or many little conscious entities. They decide that the cloud has developed highly effective means of internal communication by electromagnetic fields. Each small part of the cloud is able to send signals to other parts in select frequency bands. The imagined process is analogous to the modern Internet, but with a large fraction of the earth's population online at all times and rapidly sending messages to an ever-changing subset of recipients.

The scientists decide that because the rate of information transfer between all parts of the cloud is so high, the sub-clouds cannot be considered as having separate consciousness. Any thoughts or emotions experienced by a sub-cloud are assumed to be quickly and fully transmitted to many other sub-clouds. Thus the black cloud is deemed to contain a single global consciousness. Individual sub-clouds might be concerned with focused tasks like celestial navigation, memory storage, internal regulation of bodily functions, and so forth; however, for the cloud to behave as a single mind, the sub-clouds must orchestrate their work in concert. Neuroscientists refer to the analogous brain processes as *functional segregation* and *functional integration*. Different brain regions do different things; they are segregated. At the same time, they cooperate to yield a unified behavior and unified consciousness, so they are integrated. The question of how this can be accomplished in human brains is known as the *binding problem* in neuroscience. The idea that brain networks are *bound by synchrony* seems quite plausible when viewed in this context.

The local-global brain system of figure 6-12 might also be viewed as a fanciful representation of the black cloud's "nervous system," imagined to be composed of many sub-clouds analogous to local networks. This picture also provides an idea of how human brains may operate by means of the interactions between many subsystems. The common symbols represent *functional connections* rather than fixed connections between tissue masses. While the human brain contains more than a hundred trillion (10^{14}) "hardwired" fiber connections, figure 6-12 represents only a "time-smeared snapshot" of active functional connections that last for at least several hundred milliseconds. The functional connections may turn on and off in fractions of a second. In our fanciful black cloud, this process of switching specific interactions between sub-clouds is accomplished by tuning electromagnetic transmitter signals to match the *resonance* characteristic of the receivers, analogous to a gigantic system of broadcasting stations and TV tuners, also operating on resonance principles. In chapter 7, I propose the conjecture that *resonant interactions* between brain masses may allow for an analogous selectivity of communication between brain sub-systems, thereby addressing the binding problem of brain science.

Wait just a minute! My mental crystal ball showing pictures of future events just lit up; some readers are now thinking, "Who is this crackpot who claims that brains consist of little gremlins sending cell-phone messages to each other?" But unlike the black cloud's means of selective communication, my proposed brain resonance effects have little to do with electromagnetic fields and cell-phone antennae. Brain resonance has been established experimentally at several organizational levels; not so surprising since an uncountable number of neural network models have anticipated this result. In simple terms, resonance is just about everywhere; why not the brain?

6.7 A SHORT HISTORY OF ALPHA RHYTHMS

*Alpha rhythm*s may be defined as EEG signals with relatively large frequency components in the 8 to 13 Hz range when recorded from the human scalp. Here we spend a little extra time on this topic because of the close relationship of several kinds of alpha rhythms to local-global dynamic patterns and spatial-measurement scales. Alpha rhythms were first recorded by the German psychologist Hans Berger in 1924; the early subjects were mostly Berger's teenage children. Alpha-rhythm amplitudes on the scalp are typically in the range of 20 to 50 microvolts (μV), more than thirty thousand times lower than the voltage produced by a single AA battery. An important barrier to reliable recordings is scalp-recorded artifact of various kinds—electrical noise from the laboratory environment as well as biologic artifacts like heart, muscle, eye, and body-movement voltages. Even with the technological advances of the past century, the EEG artifact issue remains very much alive today. In short, we really want to know what's going on in the brain, not just the junk in the scalp.

Because of his doubts that he was actually recording genuine brain activity, Berger took five years to publish his first paper in 1929; his translated title is *Recording the Electrical Activity of the Human Brain from the Surface of the Head.* His findings were first met with disbelief and contempt by the German medical and scientific establishments. However, by 1934 British scientists had confirmed Berger's observations. By 1938, electroencephalography (EEG) had gained stamps of approval by many eminent scientists, leading to expanded research and use in medical diagnoses, practiced mainly in the United States, England, France, and the Soviet Union. Since that time EEG has continued to provide an amazingly robust window on the working mind.

This short history might make us stop and think about its implications for current and future findings of brain science that may or may not prove to be valid. Perhaps Berger's problems with getting his work accepted might cause us to think, "Oh, those dumb scientists; they couldn't recognize EEG as a fantastic new tool to study consciousness when it was staring them right in the face." But this attitude is unwarranted; it only looks valid with the benefit of hindsight. Whenever some new scientific finding is published, other scientists must decide whether or not to carry the work to the next step. So if one hears of some so-called breakthrough, others must make educated guesses as to whether going to the trouble and expense of new research directions is a good bet. Such decisions are based on both science and "art," meaning our intuitive feelings about the implied gambling odds.

New results that seem to violate known physical or biological laws will almost certainly be regarded as bad bets. However, the number of potential experiments that are consistent with accepted science is essentially infinite. Suppose we believe that some discovery can have profound consequences if confirmed, but the chances of successful verification are low; when should we bet on such long shots? Let's say you go to the racetrack and examine a really good-looking horse waiting in his stall for a few minutes of glory. He's a 50 to 1 shot; do you bet on this hay burner? Maybe you bet your lunch money, but probably not your year's salary. In chapter 10 we return to this scientific gambling issue in connection with certain aspects of quantum mechanics, controlled fusion, searches for extraterrestrial intelligence (SETI), and extrasensory perception (ESP).

In modern EEG work the basic question of whether or not some recorded signal actually originates in the brain is a continuing source of concern for serious scientists. Unfortunately, there have been some scientists who didn't seem to worry much about this issue; they simply declared their data to be "artifact-free." Thus, the artifact problem was simply conjured away as if some magician waved his magic wand and "poof!" it was gone. Earlier we cited studies of gamma rhythms (faster than about 30 Hz) recorded in animals and human epileptic patients. One can also record "gamma rhythms" from the scalp. It's easy; just ask the subject to clench his jaw—a flood of gamma oscillations will then be produced by scalp muscles. Good EEG scientists must assume the vigilant role of sheriffs protecting genuine brain signals from the evil artifacts. How can we tell the difference? There are several solutions, including proper use of mental tasks, checks for consistency, focus on narrow frequency band responses, and so forth. As in all of science, the devil is in the details.

6.8 ALPHA RHYTHMS: THE GOOD, THE BAD, AND THE UGLY

A posterior rhythm of approximately 4 Hz develops in babies in the first few months of age. The amplitude increases with eye closure and is believed to be the precursor of at least one of the mature alpha rhythms. Maturation of alpha rhythms is characterized by increased frequency between ages of about three and ten. Alpha rhythms, like most EEG phenomena, typically exhibit an inverse relationship between amplitude and frequency. Hyperventilation and some drugs, like alcohol, may cause reductions of alpha frequencies together with increased amplitudes. Other drugs, like barbiturates, may be associated with increased amplitude of beta activity superimposed on scalp alpha rhythms. The physiological bases for the inverse relation between amplitude and frequency and most other properties of EEG are largely unknown, although physiologically based theories have provided tentative explanations.

Alpha rhythms provide an appropriate starting point for EEG medical exams. Some initial clinical questions are, Does the patient show an alpha rhythm with eyes closed, especially over posterior scalp? Are its spatial-temporal characteristics appropriate for the patient's age? How does it react to eyes opening, hyperventilation, drowsiness, and so forth? For example, disease is often associated with pronounced differences in EEG recorded over opposite hemispheres, or with very low alpha frequencies. A resting alpha frequency lower than about 8 Hz in adults is considered abnormal in all but the very old.

The largest alpha amplitudes usually occur over electrode sites in the back of the head (near the occipital cortex), and when the experimental subject or patient is awake and relaxed with closed eyes. Many have interpreted this experimental observation as evidence that alpha is generated in the occipital cortex. Others have dismissed the alpha rhythm as a simple idling of the brain when it has nothing better to do. My interpretation of alpha is quite different, involving multiple alpha rhythms, at least one global and several local. Such rhythms provide genuine signatures of consciousness. Furthermore, the singular label "alpha rhythm" is misleading and associated with several scientific blunders, as I discussed in "Fallacies in EEG," a chapter in *Electric Fields of the Brain*.[14] My Internet search for "alpha rhythm" yielded about 1.5 million hits, but more than a little of this material is marginal science. Just to cite one common error—averaged over time, the largest alpha amplitudes do appear near occipital and parietal regions with somewhat lesser contributions from frontal regions. Many amplitude maps, based on long time averages, have been published showing "hot spots," often in bright red, over posterior regions, thereby contributing to the view of alpha as a localized phenomenon. Unfortunately, such pictures misrepresent what is in reality a much more complex process. In short, such crude time averages of complex dynamic processes of any kind can be highly misleading if not properly interpreted.

Modern studies of alpha rhythms recorded from the human scalp with high-density electrode arrays have confirmed several essential features observed in cortical recordings by the EEG pioneers of the 1940s through 1960s. The alpha band in waking humans encompasses a complex mixture of distinct phenomena determined by different reactivity to tasks, spatial distribution over the scalp, and frequency sub-band. For example, some tasks cause upper and lower band alpha amplitudes to change in opposite directions. Some of the early findings of cortical surface recordings (ECoG) have recently been rediscovered after years of scientific amnesia. These modern results have largely confirmed the following description by EEG pioneer Grey Walter:

> We have managed to check the alpha band rhythm with intra cerebral electrodes in the occipital-parietal cortex; in regions which are practically adjacent and almost congruent one finds a variety of alpha rhythms, some are blocked by opening and closing the eyes, some are not, some respond in some way to mental activity and some do not. What one can see on the scalp is a spatial average of and large number of components, and whether you see an alpha rhythm of a particular type or not depends on which component happens to be the most highly synchronized process over the largest superficial area; there are complex rhythms in everybody.[15]

Many modern studies of alpha rhythm-related signatures of mental state have been published; here I will outline one simple experiment that several of my students and colleagues carried out in collaboration with me over several years.[16] Our volunteer subjects, often one of the students or participating scientists, were fitted with an electrode cap like the one shown in figure 6-4. The subject closes his or her eyes and is told to relax during five alternating one-minute periods depicted by the white rectangles in figure 6-13. During the five intervening one-minute periods indicated by the gray rectangles, the subject performs the following mental calculation. Starting with a number supplied by the scientist, the subject progressively adds 1, 2, 3, and so on to keep a running total—for example, 7, 8, 10, 13, 17, 22, and so on up sums of several hundred. This task involves an active component (performing the sums) plus a working-memory aspect needed to remember the previous total at each step.

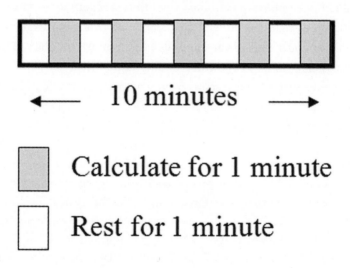

Fig. 6-13.

The subject performs this moderately difficult task with closed eyes and without body movement; both conditions tend to minimize artifact contamination of the EEG. Another feature of this experimental design is that tests for the consistency of results are easy to obtain. The ten one-minute periods indicated in figure 6-13 involve nine transitions between resting and calculating states. By recording EEG from more than one hundred scalp locations, we can look for consistent changes in the amplitudes of different rhythms, synchrony, coherence, and other kinds of dynamic patterns that will be discussed in the next chapter.

Numerous studies in the past, including Berger's original work, have found reductions in the so-called alpha rhythm during moderate to demanding mental tasks. But our studies, based on narrow-band frequencies and summarized for a "typical" subject in figure 6-14, reveal more-nuanced results. The black regions indicate frequency bands where rhythm amplitudes (averaged over all scalp sites) always increased during the five transitions between resting and calculation states, and always decreased during the four transitions between calculation and resting states. In other words, mental calculations were consistently associated with increases in upper-alpha-band (mostly 10 to 12 Hz) and theta-band (mostly 5 to 6 Hz) amplitudes.

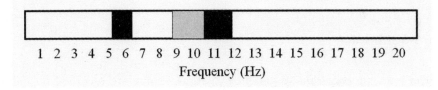

Fig. 6-14.

The gray region (8 to 10 Hz) indicates just the opposite result—lower-alpha-band amplitudes consistently decreased during mental calculations. Only these small parts of the displayed spectrum (1 to 20 Hz) yielded changes that were consistent across all nine state transitions; no repeatable changes in the delta or beta bands were observed. Of course, if we were to relax the consistency requirement, moderate statistically significant effects in the delta and beta bands would be found as well. The amplitude reductions in the lower alpha band tended to be larger than the amplitude increases in the upper alpha band. Thus, if we were to lump the entire alpha band (8 to 13 Hz) into a single category, we would find that "alpha" is reduced or "blocked" by mental activity, as many earlier studies have reported. Thus, the common phrase "alpha blocking" is, at best, misleading. These results remind us again of the intricate, detailed nature of the dynamic patterns associated with different kinds of mental activity; the simple Christmas-lights analogue fails miserably in this case.

What about the famous gamma rhythms, seemingly so important in intracranial studies of mental activity, at least in animals? In this experiment, our results in the gamma band were inconsistent and could not be interpreted in terms of brain function—we were unable to distinguish muscle and eye artifact from high-frequency brain signals. However, my guess is that LEP or ECoG electrodes would have found reliable gamma changes between resting and calculation states, but the details of such expected results are unknown. In my experience with experimental EEG, a whole lot of data gets dumped in the trash can because we can't verify that it's genuine. This general observation is certainly not confined to brain science. Again, good science should involve numerous checks and cross checks before the data are allowed to be published.

6.9 SUMMARY

The brain produces many kinds of rhythms—some are recorded on the human scalp with EEG; others are observed only with intracranial electrodes. The provocative label "rhythms" is appropriate for an *orchestra*, the brain metaphor proposed in this chapter. Brain rhythms are closely related to a broad range of mental functions, including information transfer, perception, muscle control, memory, and general states of consciousness. The structures creating these rhythms can range from single neurons to nested neural networks of different sizes, and even to globally generated phenomena; thus brain rhythms are expected to be scale-sensitive in experimental work.

Earlier chapters employed the football-fan metaphor to demonstrate states of functional localization versus states of global coherence. The magical-Christmas-lights metaphor indicated how distinct dynamic patterns of brain activity are measured at different spatial scales. Our fanciful orchestra is yet another system producing activity at multiple scales—a system of vibrating musical instruments with no conductor or conscious musicians. Orchestra rhythms are generated at several levels of organization—small-scale rhythms originate with violin strings, which stimulate a different mix of overtones produced by the violin body. The orchestra's string section provides an even higher organizational level. Similarly, brain rhythms are generated at several or perhaps many organizational levels, ranging from single neurons to the entire cerebral cortex and perhaps even the entire brain.

The frequencies at which objects like musical instruments tend to vibrate are the object's natural frequencies. The violin has four strings; each string has its own set of natural frequencies, consisting of the fundamental and its harmonics. The strings themselves produce very little sound; most of the sound is produced by the violin body, which responds selectively at its own natural frequencies, called the overtones. Resonance is this tendency to respond strongly only to certain frequencies of external stimulation—the violin body responds to oscillations of all four strings, but it responds most strongly to oscillations of the A string. Each string's set of natural frequencies, the fundamental and higher harmonics, may be adjusted by turning knobs at the end of the violin, thereby changing the tension in selected strings; tighter strings produce higher frequencies. In comparison, brain frequencies are controlled by various chemical neuromodulators.

EEG consists only of the large- to intermediate-scale brain rhythms that manage to reach the scalp; oscillations generated at smaller scales can be studied only with electrodes implanted inside the skull, either on the brain surface or inside the brain itself. One important type of oscillations consists of the alpha rhythms, a class of EEG signals with relatively large frequency components in the 8 to 13 Hz range when recorded from the human scalp. Alpha rhythms can provide a starting point for clinical EEG exams, posing questions about reaction to eyes opening, mental activity, hyperventilation, drowsiness, and so forth. Human alpha rhythms are part of a complex process, involving local networks as well as globally coherent phenomena.

Loss of normal consciousness occurs during sleep, consisting of progressively deeper stages 1 through 3, plus the rapid eye movement (REM) stage associated most strongly with dreaming. Sleep and waking are controlled by neuromodulators, special chemicals released by cells deep in the brain. The sleeping brain apparently produces an imaginary dream world nearly every night; dreams are mostly disconnected from the real world. Muscle electrical activity is suppressed during the REM stage, preventing sleepwalking in most people.

When comparing scalp-recorded EEG generated over a broad range of mental states, large amplitudes and low frequencies usually occur together. Moderate to demanding mental activity is associated with relatively low amplitudes and high frequencies. As sleep progresses from stage 1 through the deeper stages 2 and 3, EEG amplitudes increase and frequencies decrease. On the other hand, REM sleep, involving a "dream time" consciousness, consists of low amplitudes and high frequencies. Deeper anesthesia and deeper comas also occur mostly with progressively lower frequencies and high amplitudes.

Many of these results run directly counter to the metaphorical Christmas-lights brain, where deeper thinking is supposed to occur with more or brighter lights. In fact, the scientific truth is mostly the opposite— the larger the EEG amplitude, the more likely the subject is unconscious, in the states of deep sleep, anesthesia, coma, or epileptic seizure. The main reason for this outcome has to do with the local-global balance described in connection with our football-fan metaphor—a healthy awake brain must operate between the extremes of global coherence and functional localization. Almost none of this detailed information about intricate dynamic patterns is obtained with fMRI, which involves time averages over seconds or minutes. Brain rhythms are "washed out" by fMRI technology. Again, I emphasize that the various measures of brain activity provide complementary patterns of brain behavior at different spatial and temporal scales. The oscillations discussed here form only a part of a much larger "zoo" of rhythms. Many abnormal rhythms have been recorded from the cortex and scalp of epileptic patients. These include unusually large rhythms in the delta, theta, alpha, beta, and gamma bands. A classical epilepsy signature is the 3 Hz *spike and wave*, which often occurs globally over the entire scalp. Spike-and-wave traces typically occur during nonconvulsive absence seizures in which awareness is lost for seconds or minutes.

Human recordings from the cortical surface (ECoG) or from inside the brain (LEP or single neuron recordings) are limited to patients with electrodes implanted for medical reasons. In addition, intracranial recordings in living animals investigate a broad range of scientific issues. Signatures of consciousness have been discovered by employing electrophysiology at several different scales. As expected in any genuine complex system, the results are often scale-specific. For example, the dynamic patterns of neuron-firing rates in a local tissue mass are likely to look quite different than larger-scale measures of identical mental activity.

In the early 1990s, neuroscientist Wolf Singer advanced a provocative idea—different features of an animal's world are represented by distinct networks that can be functionally connected by synchronous rhythms. The catchy label "bound by synchrony" then entered the language of neuroscience. Most of the intracranial work has been focused on gamma rhythms, oscillations in the 40 Hz range; however, signatures of consciousness occur at a number of spatial scales and in multiple (but select) frequency bands. Good mental health seems to require properly tuned brains to orchestrate the music of consciousness. But healthy brain music may require multiple rhythms, measured at different spatial scales and in several distinct frequency bands.

A model brain seemingly consistent with complex dynamic processes contains distinct large-scale networks, indicating tissue that remains functionally connected for at least several hundred milliseconds. Multiscale networks are embedded in global systems that may constitute the cognitive scientist's *global workspace*, where information associated with consciousness is exchanged between subsystems. Functional connections may turn on and off in fractions of a second; resonant interactions between brain masses may allow for communication selectivity between brain subsystems, thereby addressing the binding problem of brain science.

Alpha rhythms are emphasized because of their close relationships to the central issues of local versus global dynamic patterns and the disparate measurement scales of electrophysiology. Also discussed in the same context are some differences between good and bad EEG science, for example, in the contexts of artifact and poor frequency resolution. Scalp artifact consists of noise present in the laboratory environment as well as biologic artifacts, the electrical currents produced by the heart and muscles, as well as eye and body movements.

Distinct alpha rhythms are distinguished by different responses to tasks, spatial distribution over the scalp, and frequency sub-band. Early findings of scalp and cortical surface recordings have recently been rediscovered after years of neglect. A variety of alpha rhythms occur—some are blocked by opening the eyes, some are not; some respond in some way to mental activity, and some do not. Scalp alpha rhythms result from the more highly synchronized alpha rhythms over large cortical areas. In short, humans produce complex alpha rhythms plus other rhythms of many kinds.

A simple experiment has subjects performing mental calculations, alternating with resting states, during ten successive one-minute periods. Amplitudes change consistently between resting and calculating states in select frequency bands—notably increased theta-band, increased upper-alpha-band, and decreased lower-alpha-band amplitudes occur with mental calculations. Traditional "alpha blocking" is observed when the entire alpha band is treated as a single phenomenon because amplitude reductions in the lower alpha band are larger than amplitude increases in the upper alpha band. As in many kinds of experiments, different parts of the alpha band behave differently. The main lesson gained from these example experiments is this: Many large-scale signatures of consciousness are revealed with EEG; these signatures form intricate, detailed dynamic patterns of information. Distinct patterns are even discovered at the different (but large) subscales of cortical recording as well as both high- and low-resolution EEG, as discussed in the next chapter. The simplistic Christmas-lights brain provides, at best, a severely impoverished analogue of these highly detailed dynamic patterns. In later chapters, I will argue that today's neuroscience has only begun to breach a thick veil of ignorance, possibly hiding many additional multi-scale dynamic patterns of consciousness.

Chapter 7

BRAIN SYNCHRONY, COHERENCE, AND RESONANCE

7.1 SEPARATED SYSTEMS INTERACT TO FORM LARGER NETWORKS

Brains are composed of neurons and supporting tissue; the higher organizational levels are labeled *cell assemblies*, groups of neurons that remain *functionally connected* for at least a few hundred milliseconds. Cell groups can be functionally connected in many and varied ways, as established by different experimental measures of such connections. Strictly speaking, the label "cell assembly" is more accurate than "neural network" because the latter may falsely imply electric circuit or other fixed structural analogues. By contrast, brain cell assemblies are believed to continuously form and dissolve, creating ever-changing dynamic patterns. We will, however, mostly adopt the handy tag "network," consistent with common-language use. Neural networks are believed to provide an underlying basis for mental activity, both conscious and unconscious. Chapter 8 will consider brain networks in more depth.

Here and again in chapter 8 we will look for ways in which small networks or parts of large networks can become functionally connected, but also form and dissolve quickly as mental states change. A closely related issue concerns the means with which networks can combine to form progressively larger networks, yielding nested hierarchies—networks within networks within networks, and so forth. Networks consist of smaller parts (*nodes*), which interact through *links* connecting the nodes. In other words, network nodes exchange information of some sort. A simple social network was depicted earlier in figure 1-3, where the nodes are persons, and the links represent any kind of paired social or physical interaction between the individuals in the social network.

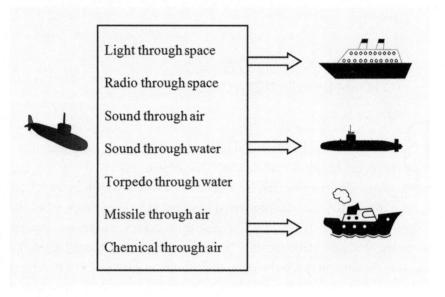

Fig. 7-1.

But before getting deeper into specific network issues, let's first lay some groundwork by considering several ways in which information is communicated in nearly any kind of system when its smaller parts (subsystems) are separated in space. That is, we wish to examine how links between subsystems can be created, with the aim of applying these general ideas to brain networks. To this end, consider figure 7-1, which provides an overview of possible means of communication between ships at sea—we may label any such information transfer as a "signal" even if the information is conveyed in the destructive manner of missiles or torpedoes. The arrows indicate communication in one direction from the submarine on the left side to the ships on the right; however, our arguments apply equally to signals in multiple directions. From the general perspective of information transfer, each signal process consists of three major parts—*sender*, *signal medium*, and *receiver*. In network language, the sender and receiver are nodes and the signal medium provides the link. For now, we ignore the issue of signal decoding by receivers.

In the first four cases, the *sender* generates either electromagnetic waves or sound waves in which the medium (space or air) fixes the speed of signal transfer. This dependence of propagation speed on medium properties is a basic characteristic of all waves that propagate in water, air, earth, musical instruments, and even the "medium" of empty space. The next two "signal processes" employ torpedoes or missiles, with their speeds determined mainly by the energy supplied, although limited by the resistive forces of the medium (water or air). Another general consideration of signal transmission is target specificity. Submarine senders can broadly target all nearby ships with light flashes or through the downwind release of chemicals. Similarly, some brain cells release neuromodulators that spread through the entire brain, potentially influencing neurons in many locations. Alternately, submarine senders can focus on specific targets using torpedoes or missiles, special kinds of "signals" in our choice of language. The torpedo and cruise missile are appropriate analogues of the action potential, which targets a specific set of cells. In both the action potential and torpedo/missile examples, energy is supplied to the moving entity along its guided pathway to the selected target.

The properties of the *medium* through which signals pass is another general aspect of information exchange. In the examples of radio and light, the relevant medium is empty space, which has physical properties that allow electric and magnetic fields to couple (join forces), thereby generating signals that propagate at the speed of light. The air normally has minimal influence on light propagation except in some cases like dense fog; thus space itself, not air, is the appropriate medium for light and radio transmission. By contrast, the relevant media for sound waves are air or water. Other media of note include the earth's crust, supporting seismic waves that carry signals from earthquake zones (senders) to buildings (receivers). Still other examples were discussed in chapter 6, where mechanical vibrators like violin strings, violin bodies, and other parts of musical instruments serve as signal media. With the notable exception of empty space, all signal media cited here are composed of molecules. But just what are molecules? Molecules are basically atomic networks. Following similar reasoning, we see that brain-tissue mass, composed partly of nested networks at multiple scales, provides a bona fide medium for large-scale signal transmission in the brain.

Receivers of signals may be humans whose visual systems are finely tuned to light; or they may be radios or cell phones finely tuned to specific input frequencies. In both cases, the fine tuning is accomplished by *resonance phenomena*, and we will have much more to say about this in chapter 8. On the other hand, the receivers might be entire ships, which are not tuned to torpedoes or missiles; however, all such "signals" sent by enemy ships are "recorded" if the target is hit. The ships themselves will not normally detect chemical signals, but humans on deck can act as special chemical receivers. Maybe they just receive harmless smells from the sending ship's galley, or maybe a chemical weapon is the source. These same general arguments may be applied to interactions between neurons and cell assemblies at different scales. In each case, we may reasonably pose basic questions about the nature of the sender, medium, and receiver.

A number of issues relevant to neural networks may be demonstrated with assistance from table 7-1. One question concerns the selectivity of receivers; every tissue mass listed as a receiver may be exposed to many kinds of signals. Yet each tissue normally remains faithful to its partner, a special kind of sender. This selectivity is achieved in several ways: In three of these cases (2, 3, 7) action potentials follow axon pathways to specific targets. Action potentials are similar to guided electromagnetic waves that propagate along power (transmission) lines or Internet cables, although their underlying physical bases differ substatially.[1] In such cases, point A is directly connected to point B; no other tissue need be involved in the information transfer. By contrast, neuromodulators (5) and neurotransmitters (6) employ chemical specificity; that is, the sender provides a "key" chemical that only fits certain "locks," the receptor molecules in the receiver. Finally, light (1) and sound (4) obtain specificity by means of resonance phenomena; that is, the receptors only respond to a narrow range of frequencies of the sender's "rhythms."

Table 7-1 lists examples of senders, media, and receivers associated with brains. Several intermediate steps are omitted for simplicity; for example, in row 2, substantial visual processing occurs between the rod and cone cells of the retina and the million or so axons of the optic nerve. Also omitted is the synaptic relay in the thalamus before axon signals reach the primary visual cortex. None of these details is important in the general context of this section, however.

Table 7–1. Signals from Senders to Receivers.

	Sender	Medium	Receiver
1	External light source	Space	Rod or cone cells
2	Rod or cone cell	Axon	Primary visual cortex
3	Primary visual cortex	Axon	Cortex at larger scales
4	Human voice	Air	Ear basilar membrane
5	Neuromodulator cell	CSF	Widespread neurons
6	Synaptic knob	CSF	Target membrane
7	Motor cortex	Axon	Muscle fibers
8	Network	Neural mass	Other networks

Row 8 indicates a more-complicated idea concerning brain signal transmission. Later I will argue that this proposed process allows functional connections to turn off and on quickly so that networks can form and dissolve on timescales in the 100 millisecond range, consistent with the changes in mental states observed in many scientific experiments as well as our daily lives. The medium for signal propagation in this case consists of tissue masses that may contain nested neural networks at multiple scales. This process is analogous to large-scale signals passing through physical media composed of atomic and molecular networks, or to information passing through some large-scale social medium like a culture composed of multiscale social networks. The senders are neural networks; the receivers may be networks of any size embedded in the networked tissue, which may be labeled a *neural mass*, as discussed below.

7.2 MASS ACTION IN THE NERVOUS SYSTEM

The title of this section is borrowed from the work of Walter Freeman,[2] a neuroscientist who championed the idea of neural-mass action since the late 1960s. While mass action in neural tissue has been slow to catch on in some subfields of brain science, the idea follows naturally from treating the brain as a genuine complex system. The basic picture is this—when the density of synaptic and action potential activity in a brain-tissue mass is high, such that a large number of neurons mutually influence each other, new larger-scale entities may emerge from the tissue. A new entity could be a cortical column of some size, as indicated earlier by table 3-1 and figures 3-6 and 3-8. Or it could be a collection of interacting columns or some other brain-tissue mass. The new large-scale entity's properties can be expected to differ substantially from those of a single neuron or simple network— just like single-neuron properties differ from the underlying molecular networks that form the neuron. As indicated in figure 7-2, atomic networks form molecules, molecular networks form neurons, neurons combine to form level 1 networks, level 1 networks combine to form level 2 networks, and so on, until the entire brain is identified as the largest network. This process may go even further—in the more-speculative later chapters, we ask if even-higher-level networks, involving multiple brains, are plausible. But, for now, we limit our discussion to the range of spatial scales between the molecular (10^{-8} cm) and the whole brain (10 cm), spanning a size factor of one billion and plenty of room for multiple nested networks.

It is argued here that the neural mass may be treated as a medium for transmission of large-scale "signals," just as we acknowledge atomic and molecular networks as bona fide signal media. But just what are these so-called signals, and how do they differ from ordinary action potentials? To address this question, consider the tissue mass shown in figure 7-3. For purposes of this discussion, we may identify this tissue as a cortical macrocolumn; however, the basic arguments apply to nearly any other brain-tissue mass containing a large number of interacting neurons.

The macrocolumn contains about a million (10^6) neurons and ten billion (10^{10}) or so synapses. The "current state" of this tissue mass might be defined by many different measures, for example, by blood-oxygen level measured with fMRI. Blood oxygen uptake is believed to provide an approximate measure of the number of action potentials fired within the tissue mass over periods of seconds to minutes. But let's consider several alternate measures of tissue state appropriate for the quick millisecond-scale changes expected in mental activity. Two such measures are the numbers of active excitatory and inhibitory synapses at some "fixed time," let's say averaged over about 5 or 10 milliseconds. In the example of figure 7-3, there are one hundred million (10^8) active excitatory synapses, but only one hundred thousand (10^5) active inhibitory synapses; thus, the macrocolumn is shown in an excited state, implying excess action potential activity compared to the surrounding tissue. In cases where excess in synaptic excitation persists for sufficiently long times, we may reasonably anticipate a moderate to strong fMRI signal from the corresponding tissue voxels because large excitatory synaptic action normally implies a large number of action potentials. But if the synaptic excitation lasts only for a very short time, no such fMRI "hot spots" need occur. Again, our experimental results always depend on measurement scale; the timescale is critical in this example.

Generally, we expect large tissue masses with excess synaptic excitation to produce increased numbers of action potentials. This implies a tendency for the macrocolumn to spread excitation via action potentials to the surrounding tissue, as well as to more-distant brain regions, as indicated by the arrows in figure 7-3. Such general large-scale spreading of excitation may occur with or without internal network action at smaller scales. *This is a large-scale phenomenon mostly divorced from the details of smaller scales.* The large-scale behavior of the tissue mass might be expressed in terms of several kinds of *fields*, including excitatory synaptic action, inhibitory synaptic action, number of action potentials fired, probability that some neuron will fire, and so on. The word "field" as used here is essentially a shorthand label for "dynamic pattern," but with more emphasis on its mathematical aspects, for example, as when employed by *field theories* of physics like electromagnetic, quantum, or gravitational field theories. In other examples, the dynamic pattern of air pressure of a sound wave is a field because air pressure has a specific value in each macroscopic air mass and at each point in time. Analogous to neural mass action, *the large-scale phenomenon of sound is largely divorced from the molecular details of smaller scales*. Normal conversations between persons consist of field exchanges, sound waves that pass through each other. But the fields still arrive at the inner ear receivers mostly undistorted, even when we talk at the same time! In many other examples, space and time vibration patterns of physical structures like musical instruments, bridges, and airplane wings are also fields. The temperature distributed through your house is a field. The weather pattern that was shown earlier in figure 6-5 reveals a bona fide "cloud field" at a fixed time. The human population density over the earth, perhaps expressed as a video of progressive maps covering the past several hundred years, is a "population field."

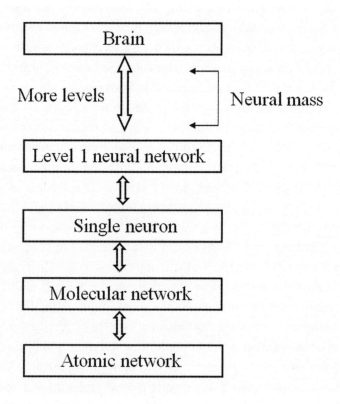

Fig. 7-2.

Cortical Tissue Mass

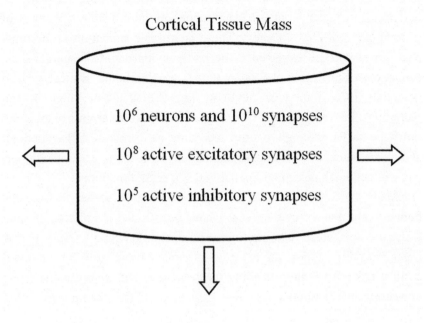

Fig. 7-3.

OK, you say, just what does this theoretical picture of fields acting in neural-tissue masses have to do with real brains? And how can fields and networks operate in the cooperative manner apparently required in healthy brains? Some preliminary answers to these questions are thus—EEG, recorded on the scalp, provides an indirect measure of synaptic action fields in the underlying cortical tissue. In other words, oscillations in the numbers of active synapses in each tissue mass result in the EEG rhythms that scientists record on the scalp. A plausible guess is that increases in the number of active inhibitory synapses and decreases in active excitatory synapses produce similar changes in scalp potentials, but this question is complicated by other issues like the actual distributions of each kind of synapse through cortical depths. In other words, the "strength" of the synaptic action in a tissue mass depends on how the synapses are distributed across the cortex.[3] But not to worry; such details are not important for our purposes. The bottom line is this—*the oscillations of synaptic action in large-scale tissue masses are reflected on the scalp as EEG rhythms.* As indicated in chapter 6, such EEG rhythms provide reliable signatures of general states of consciousness. These large-scale signatures (consciousness correlates) are expected to be partly independent of smaller-scale signatures occurring at the same time—the fMRI, the cortical surface potentials (ECoG), the local field potentials (LFP), and the single-neuron recordings. For example, the EEG can tell if you are dreaming; at the same time, the fMRI or single-neuron recording can possibly tell something about the content of your dream. Again, this picture emphasizes the importance of networks nested within networks to brain function.

Many EEG studies have recorded propagating activity across the human scalp; these *traveling waves* have been found in epilepsy, normal alpha rhythms, sleep delta rhythms, and evoked potentials.[4] Such experiments provide support for the process depicted in figure 7-3. "Excited cortical tissue," roughly meaning tissue masses with many more active excitatory than inhibitory synapses, can spread this excited level to other tissue, regardless of the activity of embedded networks. Such excitation is expected to cause increased firing of action potentials within the tissue, but here we focus on the synaptic actions, which are more closely related to the special thing we can easily measure, the EEG. The processes of large-scale spreading of excitation can apparently occur in parallel with activity of the smaller-scale nested networks that may operate within the same tissue. In chapter 8 we suggest ways in which large-scale synaptic activity may act top-down on networks, as was indicated symbolically in figure 1-4.

The EEG reveals large-scale dynamic patterns of synaptic action over the cortex; these patterns are general signatures of consciousness, providing reliable measures of different mental states, as discussed in chapter 6. But there is no apparent reason to claim that EEG is inherently any better or worse than the alternate signatures of consciousness observed at other spatial and temporal scales—the fMRI, ECoG, LFP, and single-neuron recordings. Each brain scientist typically focuses on experiments carried out over very limited ranges of organizational levels (scales), after all, no one scientist can be an expert at everything. Human nature often compels us to favor our own work over the work of colleagues, and brain science has certainly not been immune to such bias. However, if we proceed with open minds, avoiding "scale-chauvinism," each level of brain organization may be seen to reveal its own unique set of secrets. In short, *information obtained at different brain scales is complementary*—probably any good sociologist studying Earth's social networks could have told us this long ago.

7.3 WHAT ARE FUNCTIONAL CONNECTIONS?

Anatomical networks, including their nodes and links, are often considered fixed in brain science because they are assumed to change only on long timescales. However, the strength of functional neural connections (*weighted links*) can change rapidly in the so-called neural networks. (Recall that in order to keep our language in line with common use, we have adopted "network" terminology in place of the more-accurate term, "cell assembly," defined as a group of cells that act as single integrated entity, but perhaps only for a hundred milliseconds [one-tenth of a second] or so.) In this section, we retain the labels "node" and "link" but emphasize that these are not really static structures.

A link connecting any pair of nodes in a network requires information exchange between the two nodes; this exchange essentially defines the label "link." In the case of nested networks, the nodes themselves may consist of smaller-scale networks, including smaller-scale nodes and links. In complex systems like the brain, we should not be surprised by such multiscale structures. If two nodes at any scale are linked, they may be *physically connected*, but even nodes that are not connected physically can be *functionally connected*. In a simple example, suppose that two unconnected nodes A and B occupy the same network. Node A is not directly connected to node B, but both nodes are connected to node C; thus A and B are functionally connected. But the label "functionally connected" assumes nothing at all about the cause or causes of this statistical relationship.

Within the scope of our broad definition, any pair of systems (or nodes) is said to be functionally connected if any feature of one system provides (statistically significant) predictive information about the other system (or node). In other words, some activity of one system is correlated with some activity of the other system. Such connection may or may not be causal. To cite social systems as an example, note that many human activities are functionally connected if only because they share the same twenty-four-hour cycle imposed from above. Ivan lives in Moscow and Isabelle lives in New York; they don't know each other, and they have no mutual friends. They are not physically connected; however, Ivan's sleep stages are expected to be functionally connected to Isabelle's sleep stages if they follow similar local sleeping habits. The same argument applies to other physiological measures. If Ivan and Isabelle have similar eating habits, their blood sugar levels should approximately match after an eight-hour delay, reflecting the difference in time zones between their two cities.

This definition of *functionally connectivity* may, at first, seem far too broad to be useful; maybe nearly everything will be found to be functionally connected to nearly everything else if we just look closely enough at all possible predictive relationships. However, functional connectivity measures of various kinds may be expressed in terms of quantitative variables, subject to measurement when brains operate in various mental states. As such they can provide useful signatures of consciousness. EEG, in particular, reveals functional connections between synaptic action in different parts of the cortex that change reliably with various mental tasks. These functional connections cannot be due only to physical (anatomical) links, because they can turn on and off rather quickly.

When two nodes or systems are functionally connected, they are *correlated* in some way. In simple terms—when some quantity in one system increases, some quantity in the other system also tends to go up. The correlation may occur at a fixed time point, or it may involve a time lag, as in our fanciful example of Ivan and Isabelle. The blood sugar levels of these two citizens are said to *covary*, that is, to be *cross correlated*, with an eight-hour time lag. This information, however, tells us nothing about possible the cross correlation (*covariance*) at other time lags or of other features (variables) that may or may not be correlated in some way. The conventional dictum that "correlation does not imply causation" means that correlation cannot be used to infer a causal relationship between entities. Ivan's bedtimes are obviously not caused by Isabelle's earlier choice to hit the sack.

7.4 TIME DELAYS AND DYNAMIC PATTERNS

Given the general picture of many local networks embedded in a neural-mass system, one may ask about the kinds of large-scale patterns we might expect to find in scalp recordings. In particular, does EEG display synchronized activity associated with consciousness similar to the gamma-band (near 40 Hz) synchrony found in local field potential (LEP) studies discussed in chapter 6? The answer is an unequivocal yes, provided we replace the label "synchronization" with the closely related term "coherence," a type of correlation discussed in the next section. Reliable coherence patterns obtained from human scalp recordings are typically found at frequencies well below the gamma range, typically in the theta and alpha bands (4 to 13 Hz). The key to the important distinction between synchrony and coherence is the fact that action potentials carrying signals between widely separated brain regions can take significant times to arrive. Speed depends partly on axon diameter; the larger the faster. More important, speed is increased substantially by *axon myelination*, the special cells that wrap around axons. Many different white matter diseases attack the myelin wraps, leading to network timing changes.

Axon propagation delays are at least partly responsible for our internal delays of consciousness and can produce interesting consequences. Suppose you are driving on the highway at 68 mph (about 100 feet/second), closely following the car in front like the typical tailgater. During the time for a brake-light signal from the car in front to traverse your optic nerve and reach your cerebral cortex, your car has traveled only about 3 feet. However, when your foot first hits the brake pedal due to your unconscious reflex, the car has traveled about 20 feet. Your conscious awareness of the brake light does not occur until later, when the car has traveled 30 to 50 feet or so, meaning that over a short interval of awareness, you will be unaware that you have already crashed into the car ahead.

As a general rule, we should expect the dynamic patterns expressed in complex systems to be strongly influenced by time delays in information passing between elements of the system. Furthermore, these time delays are likely to be longer in larger systems. As we shall see, this implies that, other things being equal, lower frequencies may provide better signatures of consciousness in larger brains. This idea embodies important implications when the results of animal experiments are extrapolated to humans. In simple terms, information transfer between nodes takes time, and such delays impact the dynamic patterns observed, including the dominant frequencies of these patterns.

Functional connectivity with time delays is nicely demonstrated by the work of cognitive scientist Alan Gevins.[5] Gevins and his colleagues employ *event-related potentials* (ERPs), a branch of EEG research that records brain responses to external stimuli in conjunction with some task. In one such experiment, the subject is given a cue to prepare for an instruction coming one second later. The instruction is a number that informs the subject how hard to push a button. The accuracy of his response is later revealed to the subject by a computer monitor. Evoked scalp potentials are recorded over several sub-second intervals, including the interval before the presentation of the number and the interval after the subject's accuracy is revealed. One critical aspect of these kinds of experiments is the appropriate employment of so-called *high-resolution EEG* methods, meaning improved spatial resolution of brain-surface potentials, as outlined later in this chapter.

These ERP experiments reveal characteristic *covariance patterns*, that is, correlations between evoked potential waveforms recorded from different cortical regions as a function of time lag. The observed correlations between posterior and frontal regions of the scalp are typically largest at 30 to 80 millisecond time lags, reflecting signal propagation and other delays. Interestingly, covariance patterns depend on button-push accuracy. Thus, the subject's performance accuracy can be predicted in a short period just before the button is pushed when brain *preparatory networks* evidently form. Figure 7-4 shows a covariance pattern produced in preparation for an accurate response with the right hand; lines between pairs of electrode sites indicate statistically significant covariances (functional connections). The implied underlying large-scale network involves links between the frontal lobe of the left hemisphere and several occipital and parietal regions. Generally, widely separated regions of both hemispheres are typically involved in the preparatory patterns. Gevins tags his covariance patterns *shadows of thought*, with Plato's allegory of the cave in mind. Compare these results with the simple Christmas-lights brain model, where one might imagine that different mental tasks are generated by corresponding brain regions, a picture partly consistent with simple muscle commands like raising your right arm. In contrast, even moderately complex mental tasks, like pushing a button with a chosen force, attempting a jump shot, or strumming a guitar string, are closely associated with functional connections between widely spaced brain regions. In the latter cases, the Christmas-lights model must be replaced by more-realistic models employing dynamic patterns of interdependency between brain regions. Again, these patterns include functional connections that can turn on and off very quickly, essentially at the speed of thought.

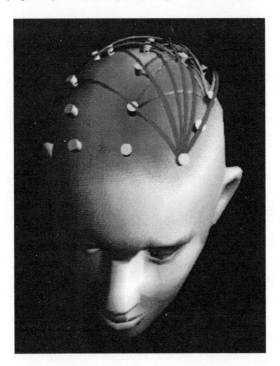

Fig. 7-4. Courtesy of Alan Gevins.

7.5 SYNCHRONY AND COHERENCE

In addition to covariance patterns, there are a number of other quantitative measures of functional connections between brain areas that may serve as reliable signatures of consciousness. One such measure is *coherence*, a quantity closely related to synchrony. The time difference between the peaks of any pair of sine waves is called the *phase difference* (as shown earlier, in figure 6-6). If a pair of signals or waveforms of any kind is composed of multiple frequency components (as in the example of figure 6-9), a distinct phase difference exists for each frequency component. Thus, two complex signals, each containing multiple frequencies, can be "in phase" (zero phase difference) at some frequencies but, over the same time interval, "out of phase" at other frequencies.

This aspect of oscillatory behavior, generally consisting of multiple frequencies, may have profound implications for brain science. It implies that separated networks can be functionally connected in some ways (at some frequencies) but, at the same time, functionally disconnected in other ways (at other frequencies). It also implies that each small-scale network can be a member of many larger networks over some time intervals, and such "alliances" can shift on and off quickly, perhaps faster than a hundred milliseconds or so.

When multiple signals are analyzed, the relative phases between signals can be of great importance, as in the case of multi-channel EEG. When two signals are lined up with zero phase offset (or nearly so) at some frequency, the signals are said to be *synchronous* at that frequency. If two signals maintain a (mostly) fixed phase relationship over a long time, the signals are said to be *coherent* at that particular frequency. Our focus on this distinction between synchrony and coherence is motivated by different kinds of functional connections between networks observed at different spatial scales— synchrony appears to be a more-important measure in small-scale systems, whereas coherence is more useful in large-scale studies.

The distinction between brain synchrony and coherence typically comes about because of significant time delays in the influences between remote networks or other systems. Suppose Professor Nancy, a social scientist, is a faculty member at "Party University." She is interested in blood alcohol responses to destructive drinking patterns. Nancy advertises on campus for experimental subjects, citing "good fun and free drinks." We assume that this imagined party school has no human subjects committee to put the kibosh on such activities.

Fig. 7-5. © Can Stock Photo Inc. / gstockstudio.

Nancy easily recruits three enthusiastic students, Jim, Steve, and Bob (depicted in figure 7-5), to join her in a neighborhood bar. During the evening, the three guys drink a beer every 16.67 minutes (1,000 seconds) while Nancy continuously monitors their blood alcohol levels. In addition to the beer, Jim has a shot of bourbon every 33.33 minutes (2,000 sec). Bob follows the same beer-drinking habits as Steve and Jim, but he also eats copious quantities of buttered popcorn along the way, thereby causing a uniform delay in his blood response to alcohol consumption. The results of this fanciful study are as follows:

- Since both Jim and Steve drink a beer every 1,000 sec, their blood alcohol levels are both *synchronous and coherent* at 0.001 Hz.
- Since Bob's blood alcohol increases are delayed, they are *not synchronous* (i.e., *asynchronous*) with Jim's and Steve's levels at 0.001 Hz. However, Bob's blood level is *coherent* with both Jim's and Steve's levels at 0.001 Hz because the time lag caused by the popcorn was consistent over their evening in the bar.
- Since Jim is the only person drinking bourbon every 2,000 seconds, his blood alcohol level is *neither synchronous nor coherent* with either Steve or Bob, at 0.0005 Hz.

What does this mean for our three barflies?
- All three guys are functionally connected at a frequency of 0.001 Hz, but they are functionally disconnected at 0.0005 Hz.
- The nature of the functional connection between Jim and Steve (no time delay) differs from their functional connections to Bob (with time delay).

Recall the waveforms shown in figure 6-6; the presence of phase lags between solid and dashed lines indicates that the sine waves are asynchronous. On the other hand, a consistent phase difference between waveforms is the mark of coherent signals. In short, "synchronous" means no phase difference between signals; "coherent" means a fixed phase difference. All synchronous signals are coherent, but the reverse is not true. Again, synchrony seems to be a more-important measure of functional connections in studies of small-scale networks, while coherence is more important in studies of large-scale networks.

7.6 HIGH-RESOLUTION EEG

One of the central messages promoted in this book is that consciousness is associated with the dynamic patterns of brain activity measured at multiple spatial and temporal scales. These dynamic patterns include different measures of functional connections between brain regions. With this background in mind, a bit of reader indulgence is requested to appreciate this short semitechnical section on so-called high-resolution EEG (HR EEG).[6] This label refers to improved spatial resolution of scalp recorded EEG by means of computer processing, but the label "high" is a bit exaggerated. In approximate terms, the resolution achieved with HR EEG is roughly two or three times better than that of unprocessed EEG.

All scientific data are recorded with limited spatial and temporal resolutions, often confined to narrow ranges as in fMRI, single-neuron recordings, LFP, ECoG, and EEG, as discussed in chapters 6 and 7. By analogy, weather channels and websites produce nice color maps of national temperature distribution at some "fixed time" (actually a time average). However, each map entry location is not a "point" but actually an estimated average temperature over some local region, perhaps a 10×10 square mile area near ground level. But many smaller hot and cold areas occur inside such large regions, forming complicated dynamic patterns that the large-scale temperature map cannot reveal. On a hot day, intermediate-scale cold regions occur inside air-conditioned homes, even smaller-scale hot regions occur near active ovens, and so forth. Even human bodies raise local small-scale temperature patterns somewhat; of course, these small-scale variations cannot be reflected in the national weather maps.

Similarly, EEG (which is recorded from the scalp) has notoriously poor spatial resolution caused by two effects: the separation distance (about one centimeter) of cortical synaptic action from scalp electrodes and the blurring influence of the intervening head tissue—mainly the cerebrospinal fluid (CSF), skull, and scalp. But, because these blurring effects are reasonably well understood, we can employ computer algorithms to sharpen the scalp patterns. This approach, known as *high-resolution EEG*, uses mathematical and computer methods to project scalp recorded potentials down to the brain surface, as summarized in figure 7-6.

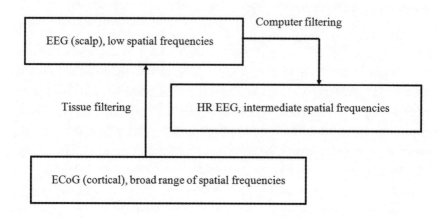

Fig. 7-6.

This computational approach involves a number of technical challenges that had to be overcome before relatively accurate estimates of cortical-surface maps could be obtained. But, for our purposes, we need not be concerned with such technical details, some of which are outlined in the notes.[7] In simple terms, the general idea of high-resolution EEG is based on spatial filtering. Imagine yourself flying over the Rocky Mountains. You see many individual peaks and valleys, but some peaks are so close together, they seem to merge. Your airplane view enjoys a spatial resolution essentially equal to the smallest distance between separate mountain peaks that you are able to discern—the lower you fly, the better your view's spatial resolution. As we have seen in chapter 6, any time signal may be broken down into individual frequency components using spectral analysis. Similarly, any spatial pattern like a mountain range may be broken down into its spatial frequency components. The lower you fly, the higher the spatial frequency components that you are able to observe, and these higher spatial frequencies reveal progressively more detail in the mountain range. Similarly, the higher the pixel density in a photograph, the more detail is revealed.

ECoG (cortical surface) patterns contain relatively broad ranges of spatial frequencies, a lot of small detail. But the higher spatial frequencies tend to be removed by tissue effects (*volume conduction*) between the cortex and the scalp. On the other hand, the very low spatial frequencies in ECoG patterns, consisting mainly of widely synchronous activity, are passed up to the scalp with less distortion and recorded as EEG. By employing clever computer algorithms, we can partly overcome this distortion, as indicated in figure 7-6. The outcomes of high-resolution EEG are spatial patterns with more-realistic balances between low and intermediate spatial frequencies, providing improved estimates of the actual cortical potential patterns.

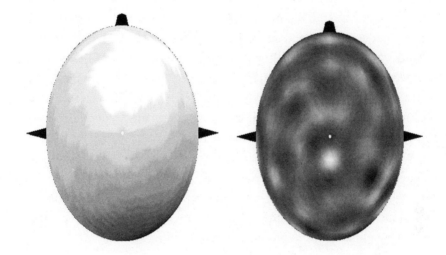

Fig. 7-7. Based on the work of Brett M. Wingeier, "A High-Resolution Study of Large-Scale Dynamic Properties of Human EEG" (PhD diss. Tulane University, April 2004).

Figure 7-7 provides an example of the spatial filtering provided by high-resolution EEG.[8] The left side is a snapshot (at a fixed time) of the scalp potential map of alpha rhythm(s) recorded from a relaxed subject with closed eyes. The map shows a broad negative potential over the frontal cortex (top) and a smooth transition to a positive peak over the posterior cortex (bottom). Because the potential distribution is smooth and gradual, we know that it consists mainly of low spatial frequencies. We also know that tissues of the head have caused the removal of the high spatial frequencies between the cortex and the scalp, and these cannot be recovered. However, if we remove the very low spatial frequencies from the map on the left side, we are left with the spatial pattern shown on the right side, a pattern that consists mainly of intermediate spatial frequencies. In summary, the head tissue removes the high spatial frequencies, and our computer algorithm removes the low spatial frequencies, leaving only the intermediate spatial frequencies, as summarized in figure 7-6.

The resulting pattern shown on the right side of figure 7-7 consists of multiple positive and negative regions of estimated synaptic source action, typically separated by several centimeters. To return to the mountain analogue, imagine a large mountain range with peaks near the bottom of the picture on the left side and a broad valley near the top. The picture on the left side reflects this overall topography but misses most of the detail. If, however, we "subtract out" (filter) the large-scale change from broad peak to broad valley on the left side, the spatially filtered picture on the right is obtained. This spatially filtered image emphasizes much more topographical detail because of its emphasis on intermediate spatial frequencies; many local peaks and valleys are evident in this snapshot. A series of such images over time reveals rhythms near 10 Hz, with regions separated by several centimeters oscillating in and out of phase. This spatial distribution looks much like a classic interference pattern of a standing wave, analogous to standing waves in physical structures like violin bodies.

7.7 FREQUENCY TAGGING

Frequency tagging is an experimental method that typically relies on *steady-state visual evoked potentials* (SSVEP); it involves subjects viewing lights or images that flicker at specific frequencies.[9] Three closely related motivations for this EEG technology are apparent. First, as we have seen in chapter 6, brains tend to do specific things in select frequency bands. In scalp recordings, mental tasks are closely associated with the theta (4 to 7 Hz) and alpha (8 to 13 Hz) bands. In addition, the higher gamma frequencies (often near 40 Hz) can provide reliable consciousness correlates in intracranial recordings. Thus, we may guess that brain responses in the narrow frequency band of a flickering light at certain low frequencies may yield reliable consciousness correlates.

The second motivation for adopting the flickering lights of SSVEP is to eliminate scalp potentials not originating in the brain, that is, to increase *signal-to-noise ratio*.[10] EEG studies of mental activity in the field of cognitive science have long been plagued by *artifact*, scalp potentials due to muscles, body movements, eye movements, EKG, and so forth. By contrast to artifact-plagued studies, the SSVEP allows brain responses to be recorded in very narrow frequency bands, often less than 0.1 Hz bracketing the light-flicker frequency. This is accomplished by spectral analysis of the SSVEP brain signals. Because artifact is mostly broadband (voltage patterns spread over many frequencies), it contributes little to the narrow band signal.

The third motivation for frequency tagging is to provide estimates of time delays between the flashing-light oscillation and each scalp electrode site based on phase differences between waveforms. Thus, each scalp location is associated with a certain delay of the fixed stimulus frequency, and such delays very according to brain state. If some task is performed while the light is flickering, we can track this delay (brain processing time) while the subject performs the task. A light flickering at a fixed frequency is required for technical reasons related to the accuracy of phase estimates, an issue unrelated to artifact.

In one SSVEP study developed by Richard Silberstein and his colleagues,[11] subjects were given a standard IQ test and then fitted with goggles producing a light flickering at 13 Hz. This upper alpha-band frequency was chosen because of its established connection to mental activity. A series of problems was shown on a computer monitor such that the problem images were clearly visible through the flickering light. With each problem presentation, subjects were presented with six possible answers; subjects responded with a "yes" or a "no," using right- or left-hand button pushes. Brain processing speed (inverse of reaction time) was positively correlated with IQ score. Hence, higher-IQ subjects showed faster brain processing in these focused studies. A special sub-group of electrode sites in frontal and central regions exhibited large inter-site coherences during the processing periods, peaking about two seconds after probe presentation. The subjects who responded more quickly to the probes were those who exhibited highest inter-site coherences at the light-flicker frequency of 13 Hz.

These findings again suggest formation of large-scale cortical networks associated with solving several different kinds of problems, for example, mental rotation of three-dimensional objects or the matrix pattern-recognition task used in an IQ test. The stronger the functional connections of these particular networks, the faster the correct answers were obtained. While demanding mental tasks occur with increased coherence between some regions, other regions show reliable reductions in coherence at the same time. In short, it appears that some "resting networks" must dissolve while the specific large-scale networks involved with each kind of problem solving emerge.

In one commercial application of this technology, experimental subjects (consumers) view video ads through the flickering light, and the resulting dynamic patterns over the scalp are analyzed to access the subject's responses to different parts of the advertisments.[12] This approach is one kind of *neuromarketing*. One aim of this procedure is to predict which parts of the ads are retained in long-term memory, as checked by written tests administered to the subjects several weeks or months later. One of the most important features of this methodology is the ability to measure variations in the delay between the stimulus and the SSVEP responses at different electrode sites over extended periods of time. This approach offers a unique window into brain function based on neural processing speed, including the influence of axon delays. When subjects view a TV commercial video, one can follow their response in real time to assess their reactions to different parts of the video. In another twist, the highly varied reactions of men and women to provocative video narratives can be assessed. For example, scenes implying male bonding are viewed quite differently by men and women. No surprise here, but my main point is that such differences can be reliably quantified with SSVEP.

Figure 7-8 provides a simplified overview of many studies of functional connections observed in different kinds of scalp recordings (EEG,

ERP, or SSVEP). In each of these categories, both unprocessed and high-resolution methods may be applied to the same data sets, providing estimates of consciousness signatures at somewhat-different spatial scales. The small black circles represent electrodes; each double arrow might indicate any one of several different measures of functional connectivity. Such measures include time (phase) delays in SSVEP studies, covariance as a function of time lag in event-related potentials (ERPs), or coherence as a function of frequency in either EEG or SSVEP studies, plus high-resolution estimates of these same data.

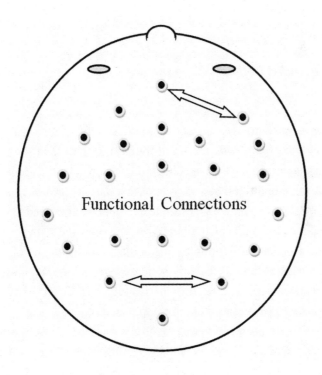

Functional Connections

Fig. 7-8.

To cite one example, a typical EEG recording might employ with 128 electrodes. There are (128 × 127)/2 or 8,128 different coherence values that may be measured at each frequency component for both unprocessed EEG and high-resolution EEG (HR EEG). With a chosen frequency resolution of 1 Hz, coherences at 1 Hz, 2 Hz, 3 Hz, and so forth may be found up to 20 Hz or more. If the corresponding HR EEG is included, this package of dynamic patterns amounts to at least $2 \times 20 \times 8,128$ or $35,512$ coherence estimates for each brain state. If the typical coherence pattern of even a single frequency component were plotted in figure 7-8, we would see a very complex network of links between electrode sites, essentially a huge plate of proverbial "spaghetti0," even if we were to eliminate all but the highest and most consistent coherences from the plot. In short, a detailed look at the dynamic patterns associated with different mental tasks reveals enormous complexity—the Christmas-lights brain model fails miserably in this case.

7.8 BINOCULAR RIVALRY

New studies appear in the scientific literature on a regular basis aiming to compare dynamic patterns produced by brain activity during some cognitive task (A) with the brain patterns of a different task (B) or of a resting state (R). These studies often reveal specific brain regions in which observed patterns show reliable differences between tasks; that is, *dynamic patterns are strongly modulated by the task.* A simple example was outlined in chapter 6, in which states of mental calculation (A) were compared with resting states (R). During calculation periods in this subject, contributions of theta and upper alpha frequencies consistently increased while lower alpha-frequency components decreased compared to the resting state. Other reliable indicators in these kinds of studies include complicated (but consistent) coherence patterns over the scalp, as discussed in the last section. To briefly summarize this complex picture—when compared to resting states (R), some coherence estimates consistently increase and some consistently decrease when subjects are performing specific mental tasks (A).

These kinds of studies provide limited signatures of consciousness, in the sense that they compare one conscious state (A) to another conscious state (B or R). Several experiments discussed in chapter 5 avoided this limitation with clever experimental designs using visual images. By employing masking strategies, forced inattention, or brief image exposure, each picture presented to the subjects can be made to vanish from consciousness. With appropriate experimental designs, pictures are perceived by subjects about half the time. In this manner, consciousness is manipulated—the subject reports seeing the image in some cases (conscious state A) but not in others (unconscious state U). Patterns associated with consciousness are then compared to unconscious patterns.

Here we describe an extension of the frequency-tagging method, designed to reveal pattern differences between conscious and unconscious perceptions. These experiments are performed under conditions of *binocular rivalry*, in which experimental subjects are presented with separate images to the right and left eyes.[13] A subject views a different image with each eye but consciously perceives only one image at a time; the image switches its perceptual dominance every few seconds. For example, the subject may be presented with a red vertical grating to the left eye and a blue horizontal grating to his right eye. Although both stimuli are presented all the time, the subject mostly reports seeing only one image or the other. The subject's visual system receives signals from both images, but somehow only one image at a time rises to the conscious level.

One such experimental variation of frequency tagging carried out by Ramesh Srinivasan and colleagues is depicted in figure 7-9. Special goggles are used to present an 8 Hz flickering blue light to the subject's left eye. At the same time, a 10 Hz flickering red light is presented to the subject's right eye. The images are also distinguished by spatial patterns, horizontal lines in the blue images and vertical lines in the red images. The subject indicates periods when he perceives only a blue or a red image with response switches; typically either purely red or purely blue images are perceived 80 percent of the time. Evoked brain responses (SSVEP) are analyzed with the usual spectral methods in the 8 and 10 Hz bands.

For brain responses in the 8 Hz band, blue perception states are known as *perceptual dominance periods* and red states as *non-dominance periods*, each typically lasting a few seconds. For brain responses in the 10 Hz band, the opposite occurs; red states correspond to perceptual dominance and blue states to perceptual non-dominance. In other words, the 10 Hz coherence plots looks much like those in figure 7-9, but with the labels "subject perceives red (blue) light" interchanged.

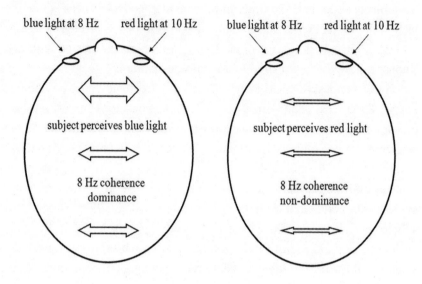

Fig. 7-9.

Similar to the other SSVEP studies cited above, scalp potentials are recorded and coherence patterns over the scalp are determined for both the perceptual dominance and non-dominance states (at both the 8 and the 10 Hz stimulus frequencies). As indicated by the widths of the arrows in figure 7-9 (greatly simplified), perceptual dominance periods are associated with increased coupling between cortical regions as measured by coherence at the frequency of the dominant color. The strongest changes tend to occur in frontal regions. In short, *a more-integrated scalp coherence pattern at each frequency occurs with conscious perception of the stimulus flickering at the matching frequency.*

These studies of frequency tagging, including binocular rivalry methods, indicate again that specific mental tasks and conscious perception typically occur with increased coherence between specific regions in some frequency bands and decreased coherence in other bands, consistent with the formation of (possibly) overlapping networks. These data address the so-called binding problem of brain science—how is the unity of conscious perception created by distinct neural systems processing different functions? Conscious perception, in these experiments, occurs with enhanced dynamic "binding" (functional connectivity) of the brain hemispheres within narrow frequency bands. At the same time, dynamic processes occurring in other frequency bands can remain "unbound," allowing similar parts of the brain to engage in independent actions. *Again, healthy brains seem to operate between the extremes of full network isolation and global coherence, and this may be accomplished in part through selective coherence in distinct frequency bands.*

7.9 MULTISCALE SYNCHRONY AND COHERENCE

The critical importance of brain rhythms to distinct mental states in particular, and consciousness in general, has long been established. Brain rhythms may be generated at multiple spatial scales, including single neurons, large-scale tissue masses, and almost anything in between. Each network may produce rhythms with its own characteristic set of natural frequencies, analogous to the different rhythms produced by different musical instruments. An important issue concerns the ways that networks interact across spatial scales, the apparent top-down and bottom-up interactions generally expected in complex systems like the brain. The relationship between the binding problem and brain resonance will be considered further in chapter 8.

The studies of the event-related potential (ERP) cited in chapter 5 reveal an important consciousness signature about 300 milliseconds after presentation of visual stimulus. This relatively large signal, called the P300, is believed to reflect synchronous cortical activity over perhaps 200 milliseconds or so; it may result at least partly from alpha-band synchrony in the cortex. Intracortical recordings in epilepsy patients reveal bursts of gamma rhythms following appearance of the P300. However, when the stimulus is not perceived, P300 is reported absent in some studies, or perhaps just reduced in amplitude; and the following gamma rhythms die out quickly. In short, consciousness of the stimulus occurs after 300 milliseconds, perhaps accompanied by a brief period of large-scale synchrony, and apparently followed by sustained gamma-rhythm activity. These experiments raise the question of possible top-down, large-scale, low-frequency influences on the smaller-scale high-frequency gamma rhythms. We will return to this issue in chapter 8.

Chapter 6 noted that intracranial studies of local field potentials (LFPs) suggest that different features of an animal's world are represented by distinct small-scale networks that can be functionally connected by synchronous gamma rhythms. In this chapter we show that large-scale networks produce their own consciousness signatures in the form of coherence patterns. That is, the existence of large-scale networks is implied by scalp recordings, including spontaneous EEG and frequency tagging (using SSVEP), the latter sometimes employing binocular rivalry in the experimental design to distinguish the different kinds of functional connections (covariance or coherence) associated with different conscious perceptions.

A simple overview of this general picture of multiscale synchrony and coherence is indicated in figure 7-10. The nodes of the imagined large-scale network, represented by the seven-sided symbols with connecting links, indicate neural masses at the (approximate) 5 to 6 cm scale of EEG or the 2 to 3 cm scale of high-resolution EEG. We postulate that smaller-scale networks may easily operate within each large tissue mass, as indicated by the enlarged symbol at the lower right; gamma synchrony may occur within this small-scale tissue. The large-scale activity is not generally synchronous, probably because of the longer time delays in information transfer between the more widely separated nodes, but the large-scale system can produce reliable coherent patterns associated with mental state. In essence, *coherence is essentially synchrony with significant time delays.*

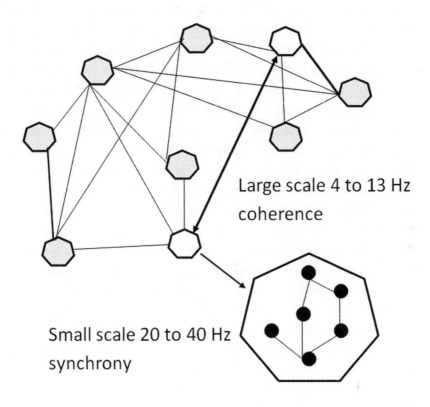

Large scale 4 to 13 Hz coherence

Small scale 20 to 40 Hz synchrony

Fig. 7-10.

7.10 SUMMARY

Brain networks continuously form and dissolve, creating ever-changing dynamic patterns that depend on internal and external information transfer. Such transmission process must include at least three features—*sender*, *signal medium*, and *receiver*. In networks, the sender and receiver are normally network nodes, and the signal medium provides the link. Receivers may be widely distributed or highly specific; brain link specificity is achieved by targeted axon connections, special chemical receptors, or resonance phenomena.

Brain network nodes may develop at multiple scales, possibly ranging from molecular levels to single neurons to networks to tissue masses containing millions of neurons. When the density of synaptic and action potential activity in a tissue mass is high, the neural mass, a new larger-scale entity may emerge within the tissue. The medium for signal propagation consists of tissue that may itself contain smaller-scale nested neural networks. Large tissue masses with excess synaptic excitation are expected to produce large numbers of action potentials, causing a local neural mass to spread its excitation to other tissue. The neural mass then essentially becomes a large-scale node in a very large-scale (perhaps global) network. The oscillations of synaptic action in large-scale tissue masses are recorded on the scalp as EEG rhythms.

Any pair of systems (or nodes) is said to be "functionally connected" if any feature of one system provides statistically predictive information about the other system (or node). Experimental studies of event-related potentials reveal characteristic *covariance patterns* associated with mental tasks. These are patterns of correlations between evoked potential waveforms recorded from different scalp regions as a function of time lag, demonstrating one particular brand of functional connectivity.

Another important measure of functional connectivity is coherence—these are (squared) correlation coefficients expressed as functions of the frequencies of various EEG brain rhythms. Coherence is a natural extension of synchrony when substantial time delays occur in information transfer between nodes or systems. EEG reveals large-scale coherent activity associated with consciousness in addition to the small-scale gamma-band synchrony found in local field potentials (LEP). High-resolution EEG provides complementary patterns at somewhat-smaller scales.

Frequency tagging employs steady-state visual evoked potentials (SSVEP), a technique where subjects view lights or images flickering at specific frequencies. These studies suggest formation of large-scale cortical networks associated with solving several different kinds of mental tasks, for example, mental rotation of three-dimensional objects or matrix pattern recognition. In addition, coherence patterns associated with consciousness may be compared with unconscious patterns by employing binocular rivalry, in which experimental subjects are presented with separate images to the right and left eyes. Subjects consciously perceive only one image at a time, but the image switches its perceptual dominance every few seconds. Conscious perception, in these experiments, may occur with enhanced coherence between brain hemispheres within narrow frequency bands. At the same time, coherence in other frequency bands can remain small (incoherent), allowing parts of the brain to engage in independent actions. Again, healthy brains seem to operate between the extremes of full functional segregation and global coherence.

Brain rhythms are generated at multiple spatial scales. Networks formed at different scales may produce rhythms with their own characteristic set of natural frequencies. Consciousness is associated with both small-scale, high-frequency (gamma) synchrony and large-scale, low-frequency (theta and alpha) coherence. In the next chapter we will see how networks and fields can partner to produce distinct dynamic patterns that manifest at multiple interacting scales. Again, social networks embedded in a culture provide us with a compelling analogue.

Chapter 8

NETWORKS OF THE BRAIN

8.1 NETWORKS, MODELS, AND THEORIES

Thus far, I have repeatedly emphasized the nested feature of brain networks, the complex subsystems engaged in both bottom-up and top-down interactions across spatial scales that produce dynamic patterns. However, most of the details are missing from this narrative. The more-focused questions concerning the detailed nature of these interactions are largely unanswered by today's science. Despite this general ignorance, we can productively recruit both experimental evidence and theoretical models to provide tentative answers. That is, these data facilitate plausible ideas about how such interactions across scales might actually occur, and how they influence or create distinct mental states. While the conceptual picture that I offer seems fully consistent with contemporary neuroscience, many of the details are uncertain.

A basic question in the cognitive studies employing brain sensory input concerns the issue of localized versus distributed influences. That is, do observed neural responses in the sensory cortex mainly reflect the actions of the *local* sensory network? Or are such responses caused by interactions between this local network and more widespread networks? Brain tissue is highly interconnected, consisting of an enormously complex system of nested networks. Thus, even when signals are recorded from a small number of neurons or even a single neuron, we generally expect the observed dynamic patterns to occur as a result of some combination of local network activity and more widespread (perhaps top-down) neural activity. The latter is anticipated because neural responses depend on the general level of arousal, which is controlled by neuromodulators, and attention or goal-oriented mental conditions that guide sensory information processing. These top-down effects on sensory neurons bias neural output and affect behavioral outcomes such as stimulus detection, short-term memory, and reaction time. This chapter focuses on the role of nested networks in the context of combined bottom-up and top-down mental processing, the "circular causality of consciousness."

The title of this chapter mirrors that of Olaf Sporns's recent book, *Networks of the Brain*—we make use of this material and its implications plus several review papers by others.[1] Sporns is one of the scientists credited with originating the term *connectome*, meaning a comprehensive map of neural connections within the brain, essentially its wiring diagram and a hot topic in brain science these days. Brain networks can be defined at several (or perhaps many) different spatial scales, typically chosen to match the spatial resolutions achieved in various brain imaging methods. These scales are often categorized as the *micro-scale*, *meso-scale*, and *macro-scale*; however, there is no good reason to limit our discussion to just three scales. We could, for example, label a single neuron as a node and member of a *level 0 network*, with progressively larger-scale networks called *level 1*, *level 2*, and so forth up to the *global* or entire brain scale. The intermediate (meso-) scale networks are based on cortical areas of various sizes serving as network nodes.

Imagine a map of the world showing all cities with populations greater than one hundred thousand, displaying the locations of something like five thousand cities, representing macro-scale nodes in this network. The dense macro-scale links connecting these nodes might consist of highway, train, ship, and airline connections. But, each city-sized macro-node contains meso-scale nodes (homes) and meso-scale links (city streets); the corresponding brain's meso-scale nodes might be cortical columns of some intermediate scale. This fractal-like structure extends to progressively smaller scales such that the micro-scale nodes of individual persons are then analogous to the single-neuron level. Micro-scale links in the world map might include walking, bicycling, horseback-riding, and even elephant trails, anything that allows human transport or information exchange between persons serving as micro-scale nodes. An alternate analogue might be the

World Wide Web, with its websites defined as nodes; however, website links are ever changing.

Given the enormous complexity of brain connections, scientists are typically faced with the task of representing genuine networks with much simpler, idealized model networks; such models facilitate good experimental designs to support future studies. We will frequently adopt this practical approach. Many scientists, myself included, often adopt careless language by mixing up the scientific categories *model* and *theory*. Much of this book is based on a *conceptual framework* consisting of nested cortical networks embedded in a global environment that includes large-scale activity, including regional networks and synaptic action fields. By itself, this framework is neither theory nor model and is unlikely ever to be proved wrong because it is based on well-established neurophysiology. Rather, an essential question is whether the proposed framework is scientifically useful; here I claim the answer is yes.

For physical scientists at least, a "theory" is normally expressible in mathematical language and makes predictions in a broad class of experiments. In this view, most so-called cognitive theories are better described as models or conceptual frameworks. A model is more restricted to specific systems and carries less status. Models often employ *Aristotelian idealizations* where all properties of the genuine system that are deemed irrelevant to the problem at hand are neglected. For example, engineers model a spacecraft's trajectory as if all of its mass is concentrated at a point called the *center of mass*. This is based on a well-established principle showing that such modeling approach is accurate. All of us model the behavior of other humans on a regular basis. We may mentally model the behaviors of priests performing on Sunday mornings or politicians being interviewed on television, all the while knowing that such models may fail completely when the same persons act in different environments.

Models also employ *Galilean idealizations* in which *essential* properties of genuine systems are deliberately neglected in order to create models simple enough to be useful. Galileo modeled falling bodies with no air resistance even though he lacked the technology to make the air go away. Modern sociological, economic, and ecological models often idealize populations and networks as existing in isolation; this approximation may or may not prove useful. Once such oversimplified models are developed, external influences may be imposed on the simple models. More-accurate models may then evolve progressively from simpler models. Weather systems provide one such example—weather models are developed progressively by integrating the theory of fluid mechanics with new experimental data.

A common scientific strategy involves a series of steps—creation of idealized (Galilean) model networks, testing the models against experiments, improving or replacing the models, and repeating the process. Science is then advanced through a happy marriage of theories or models to experimental findings. Ideally, these marriages involve healthy relationships that become progressively more intimate as time goes on. In this manner "truth" is approached in a series of successive approximations, but at no time can full and complete accuracy be claimed.

As discussed earlier, the philosopher Karl Popper is known for his proposition that *falsifiability* is the critical test that distinguishes genuine science from non-science. In this view, any claim or model that is impossible to disprove is outside the scientific realm. While much can be said in support of this idea, its application to real-world models and theory is often ambiguous. Modern philosophers and scientists vary somewhat in their opinions about the validity or practicality of falsifiability tests, but we are safe if we call it a potentially useful, but imperfect, measure.

Many mathematical brain models run counter to Popper's proposition by adopting variables, perhaps with vague labels like *activity*, having no apparent connection to any genuine experiment. Typically, these models are not right; they are not even wrong. Even if the aim is more focused, say, to model electrical events, the modeler may appear unaware of the issues of spatial and temporal scale, perhaps avoiding critical issues of sensor size and location in any related experiment. If the "activity" can never be measured, the model cannot be falsified; it's probably just computation, not science. Over my scientific career, I have often complained about "scale chauvinism," a malady inflicting scientists holding divergent persuasions about the importance of their own data relative to data recorded at other spatial and temporal scales. In contrast to such parochial positions, this book aims for a balanced viewpoint in which data collected at all scales are valued.

8.2 NETWORKS VERSUS CONTINUOUS MEDIA

We have repeatedly cited the actions of small- and intermediate-scale neural networks that are embedded in the large-scale (global) brain environment. In chapter 7, the general idea of neural mass was introduced as a means to describe the large-scale data obtained from the scalp or brain surface. But the distinction between these large-scale synaptic action fields, estimated with EEG or ECoG, and large-scale networks (cell assemblies) was left a bit vague. Here we argue that the network and field models can be fully complementary, providing useful, alternate ways of describing the same or similar phenomena.

Modeling some medium (usually matter) as a continuum assumes that the medium completely fills the space it occupies. This general "mass action approach" is known formally in a broad class of physical systems as *continuum mechanics*. This approach ignores the fact that matter is actually made of discrete networks of molecules, which in turn are smaller-scale networks of atoms. Matter is not really continuous; however, when treating problems involving length scales that are much greater than intermolecular distances, such models can be highly accurate. Continuum mechanics often deals with the physical properties of solids or fluids, but it is also applied to empty space, as in the examples of electromagnetic fields and general relativity. Modern physics tells us that the vacuum of space-time is actually a bona fide medium that carries electromagnetic energy; furthermore, even the "shape" of space-time itself is distorted locally by the presence of mass.

In each of the many physical phenomena modeled with continuum mechanics, fundamental physical laws like the conservation of mass or energy are expressed in terms of *fields*, which are essentially mathematical expressions of dynamic patterns. In chapter 10, we will see that quantum mechanics is also a continuum theory, even though it often deals with discrete entities. The field followed in quantum mechanics is not "physical" in the usual sense, rather it is an *information field* known as the quantum wavefunction.

String Networks

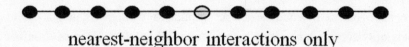

nearest-neighbor interactions only

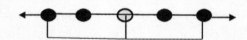

nearest and next nearest-neighbor interactions

Fig. 8-1.

Given our emphasis on complementary neural network and neural mass-action fields, it is useful to consider the relationship between these two kinds of models in the context of our proposed orchestra analogue. The violin string is not normally called a network; however, as indicated in the upper part of figure 8-1, it may be treated as a network of little balls (masses) linked to together by tiny springs that model string tension. When disturbed, the balls and springs oscillate up and down in a manner similar to the actual string. Both the string continuum and its representative network have their own separate sets of natural frequencies, each set consisting of a *fundamental frequency* and its *overtones*. However, when the number of balls is large and their spacing small, the network's natural frequencies closely approximate the natural frequencies of the violin string, provided the ball mass and spring properties match that of the string.[2] For most practical purposes, the two systems are essentially identical. In a similar manner, the violin body may be represented by a three-dimensional network used to calculate its natural frequencies with computer methods.

In the one-dimensional network shown in the upper part of figure 8-1, each ball is only linked directly to the two balls on its sides; such links are labeled *nearest-neighbor interactions*. By contrast, the lower figure shows a part of a network in which both *nearest and next nearest-neighbor links* occur (only links to the central white ball are shown). If remote balls were to be linked, we would label these links *non-local interactions*. I remind readers that this term differs from the label *nonlocal* (no hyphen), which refers to the faster-than-light-speed influences to be discussed in chapter 10. This book deals with many overlapping scientific fields, each employing its own jargon; here we will try to stick with conventional language as much as possible.

The change in connectivity structure away from exclusive nearest-neighbor interactions changes the system's natural frequencies, typically "compressing" the frequency range so that the overtone frequencies are closer to the fundamental frequency.[3] In summary, real systems can often be modeled either as networks or as continuous media, and the two views are often compatible. However, network models apply to a much larger class of systems, typically with an abundance of non-local links, which are expected to add to system complexity.

8.3 NETWORKS AND GRAPH THEORY

We have argued that the brain may be modeled productively as a complex system, consisting of nested networks in which mental states appear to emerge from interactions within and across multiple scales. This conceptual framework suggests judicious application of ideas drawn from *complex network theory*, which in the past fifteen years or so has dramatically expanded across disparate scientific fields ranging from the social sciences to physics and biology. The study of complex networks draws its ideas from physics, mathematics, computer science, and the social sciences to provide a powerful conceptual framework supporting brain study.

A fundamental mathematical basis for network analysis is provided by *graph theory*,[4] a branch of mathematics in which the labels *graph, vertex,* and *edge* correspond to the more-scientific terms *network*, *node*, and *link*, respectively. Adoption of such distinct mathematical language emphasizes that abstract discoveries about graphs are independent of whatever genuine scientific systems the graphs might represent. In this manner, one can search for general mathematical principles that apply to a broad class of systems, physical, biological, and so forth. Similarly, abstract mathematical models of resonance phenomena can provide us with general concepts that apply across many scientific fields, ideas that may qualify as "brain-friendly," as discussed later in this chapter. For this reason, the separate language of graph theory is appropriate, supporting the actual ideas that engage mathematician's brains. But here we stick with the non-mathematical scientific language, which will prove more intuitively pleasing to most readers.

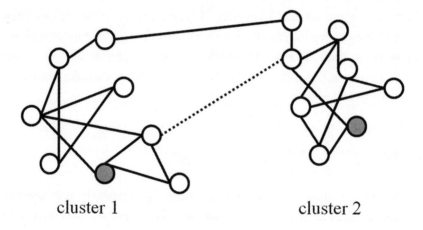

cluster 1   cluster 2

Fig. 8-2.

The network shown in figure 8-2 consists of sixteen nodes and either twenty-one or twenty-two links, depending on the presence of the dashed line. This network might be a molecular model where the nodes and connecting links represent atoms and chemical bonds, respectively. Or it might represent a social system with links indicating friendships, one-time contacts, or any kind of paired human interactions. The *path length* between any two nodes is defined as the shortest path connecting the nodes, expressed as the number of links. Thus, this path measure is does not normally correspond to an actual physical distance. When averaged over all node pairs, the *network path length* is obtained. Networks are also characterized by their connection density, expressed as the *clustering coefficient*, essentially the fraction of all possible links that actually occur; clustering coefficients always fall between zero and one.

A *small-world network* often consists of local node clusters with dense internal interconnecting links like those shown in figure 8-2, along with sparse links between separate clusters. The label "small-world" stems from the purported *six degrees of separation* between any two persons in the world. You have probably never heard of the rice farmer Mr. Long Wu who lives in the Chinese countryside; however, you may know someone who knows someone, and so forth, who knows Mr. Wu in only six links. I might boast of being only two or three links from Einstein, but only if I cheat a bit by assuming links of the past are still counted today.

Clusters (*modules*) have relatively dense internal connections but sparse connections between different clusters, as shown by the two clusters in figure 8-2. Small-world networks often have special nodes with large numbers of links (*hubs*), in the same sense that Atlanta is the hub for Delta Airlines. We anticipate correctly that increased clustering generally leads to shorter path lengths, but an interesting property of networks is how sharply this reduction in path length occurs. The addition of a small number of links between clusters can dramatically reduce path lengths even though the clustering coefficient exhibits only small increases. *The influence of non-local interactions is typically an important feature of small-world networks.* Neuroscientists who study mental processing are especially interested in small-world networks for several closely related reasons—their association with sensory information transfer between remote cortical locations, the prevalence of white matter disease, the brain binding problem, and other more-general contributions to brain complexity and consciousness.

To demonstrate small-world effects, note that the two grey nodes in figure 8-2 have a path length of six links before addition of the new (dashed) connection. After this new link is added, the grey node pair's path length is reduced to three. For the whole graph, addition of the new connection only increases the clustering coefficient from 0.175 to 0.183 (+4.6 percent). By contrast, the graph path length (average over all node pairs) falls by a much larger fraction, from 3.85 to 2.93 (−24 percent). Interactions in social networks, in which part of the population carries a communicable disease, provide practical examples. If the nodes are persons and the clusters are relatively isolated cities, disease spreads slowly or not at all between cities. But if travel volume increases beyond a critical point, sharp increases in intercity disease spread is expected. Ships carrying rats between Asia and Europe apparently provided such a critical link to create a small-world network facilitating spread of the Black Plague around the year 1400.

The mathematics of graph theory is relatively clean and well-defined; otherwise it wouldn't be called *mathematics*! But its application to brain anatomy and physiology involves many messy, non-mathematical issues. A basic question concerns spatial scale—which anatomical structures should serve as the appropriate nodes? While the single neuron provides one likely choice, cortical columns at several scales are also proposed as basic functional units of cortex that serve as nodes. As indicated in chapter 3, each cortical pyramidal cell sends local axons to the limits of its macrocolumn boundary, which encloses tissue containing about a million neurons. We don't know the actual fraction of local cells targeted by the central cell, but it is probably much less than one percent of neurons in the same macrocolumn. In addition to these intracortical axons, about a million axons from the macrocolumn provide non-local input to distant cortico-cortical columns. It appears that the cortical/white matter system fits nicely into the category of small-world network, apparently over a broad range of spatial scales. At large scales, the abundance of cortico-cortical fibers connecting remote cortical regions is apparently mostly responsible for short path lengths. Short path lengths generally suggest highly integrated dynamic behavior, but in the cerebral cortex, such functional integration is mitigated by inhibitory synaptic action.

8.4 NETWORK CAVEATS AND LIMITS TO PROGRESS

In today's scientific climate, neuroscientists rightly express enthusiasm for the discovery of more and more about the brain's wiring diagram, the so-called *connectome*. However, several important caveats should be kept in mind. One conceptual trap to avoid is a false notion that even complete knowledge of the connectome would, by itself, imply anything close to "understanding" brain operations. A more-realistic view is that brain networks with similar wiring can operate in many different ways; anatomical connectivity is just one of many issues involving brain complexity. Here we mention just a few of the areas where current scientific understanding is in very early stages.

One network class involves weighted links, accounting for different interaction strengths between pairs of nodes. Axons provide robust directed links, but often with substantial propagation delays. Synaptic plasticity during learning is believed to depend on the temporal coincidence of firing among multiple synaptic inputs to any neuron. As the old saying goes, "neurons that fire together wire together," a sentiment that might also apply to human relationships. Thus, link strengths or "effective links" may differ substantially from anatomical links. In particular, myelin controls action potential speeds and the synchrony of traffic between distant cortical regions. One implication of this feature is that myelin, along with the more well-known kinds of plasticity, appears to be critical for healthy mental states. Consistent with this picture, a broad range of psychiatric and other disorders has recently been associated with myelin defects in the white matter.

Cortical connectivity at large scales is dominated by white matter axons, especially the cortico-cortical axons, which outnumber thalamo-cortical and callosal axons by perhaps fifty to one in humans, as outlined in chapter 3. As discussed in chapter 5, axon connectivity may be estimated from postmortem dissected tissue or in living humans with *diffusion tensor imaging* (DTI). Diameter histograms of human white matter axons peak in the one micrometer range; that is, about one thousand times smaller than the one-millimeter resolution of DTI. Thus, we conclude that comprehensive mapping of axon connectivity at multiple mesoscopic and macroscopic scales may be many years away.

A broad range of *temporal* scales also characterizes human cognitive functions. For example, patterns of neuronal connections change with learning and memory by means of various kinds of plasticity on both short (seconds to minutes) and long (hours to days to months) timescales. Furthermore, the neural structure in any given spatial scale is organized into clusters or *modules*, the anatomically or functionally defined cortical areas believed to form the bases for mental functions. In the spatial domain, anatomical modules can occur as cortical minicolumns, macrocolumns, and so forth. Within modular structures, various brain areas play distinct roles, as hubs of high connectivity or as nodes of lesser importance.

In scientific fields, emergence is a process whereby larger-scale entities arise through interactions among smaller or simpler entities that themselves do not exhibit the large-scale properties. For example, life is normally perceived as an emergent property of the interacting molecules of biochemistry, which, in turn, reflect interactions among elementary particles. Higher brain functions may then be pictured as the next hierarchical step up from life. That is, most scientists view human behavior, various mental functions, and consciousness itself as properties that somehow emerge from the underlying brain networks. An essential property of brains, as opposed to some simpler versions of complex systems, is that emergent phenomena can act on lower levels, causing small-scale changes through downward causation. The top-down actions of animal brains on behavior are essentially defining features of the label "brain." Human *free will* seems to qualify as a pre-eminent example of top-down actions of large-scale systems on smaller-scale behavioral systems. Later, we carry this idea even further by considering the top-down influence of social systems on individual brains.

Despite the utility of various brain models, the investigation of links between disparate systems faces formidable obstacles, especially at the smaller-scale levels of description. Neuroscience is awash with detailed experimental data produced by a huge range of experimental methods, all measuring different aspects of the same system. The empirical compatibility of these data can be quite tenuous—it is often difficult to determine whether identical phenomenon are being measured, but just with different tools, or whether the data represent different phenomena altogether. In summary, we must ask what types of emergence characterize each brain subsystem? Do different kinds of emergence occur between different levels of the multiscale system? When are interactions across scales simply accidental correlations, and when are they causal? These are only a few of the trillion-dollar questions challenging brain scientists.

8.5 VISUAL PERCEPTION NETWORKS

The task of obtaining even a very minimal understanding of genuine brain networks presents a formidable challenge likely to engage generations of scientists for years to come. In this section, we attempt to bypass most of the attendant complexity by employing a relatively simple network model of image recognition and processing. This model serves to introduce several general ideas that correspond in some approximate ways to real brain networks. That is, the proposed model appears to be "brain friendly," in the sense of being consistent with a number of genuine experiments and not obviously in violation of any physical or biological laws. As in the case of all simple brain models, we should not expect this visual perception model to be accurate in any absolute sense; rather, our goal here is a modest one— providing useful analogues than can guide experimental designs.

The human visual system carries out a number of complex tasks, including the absorption of light, identification of visual objects, estimation of distances between objects, and guidance of the body's movements through the observed environment. The eye produces images of the visual world on the retina, which serves much the same function as the film in a camera. The retina consists of a large number of photo-receptor cells containing special protein molecules that absorb light and cause cells to fire. These proteins are selectively sensitive to narrow frequency bands of light, based on a physical process originating with *quantum resonance*, a topic that we will return to in chapter 10.

Substantial information processing occurs in neurons of the retina and several "upstream" cell layers. More than one hundred million photoreceptors absorb light; yet "only" about one million axons form the optic nerve, which transmits information to the brain—mostly through the thalamus, a relay structure that sits astride the brainstem. Visual information is passed from the thalamus to the *primary visual cortex* (labeled V1 by neuroscientists), located near the back of both brain hemispheres. Visual signals then follow two major pathways in the upper and lower parts of the brain (known as the *dorsal and ventral streams*). The secondary visual areas (V2, V3, and so forth), receive signals from V1 and then process and interpret the visual input. The conscious experience of processing visual information, which generally involves several or perhaps many back-and-forth interactions between different cortical areas, is known as *visual perception.*

To focus on one example, *face perception* refers to the interpretation of faces as a special kind of visual image; facial recognition is critically important for most human social interactions. Even one-year-old infants are able to recognize facial expressions as social cues representing another person's feelings. Face perception is a complex process involving extensive brain regions; however, some areas seem particularly important. Brain imaging studies often point to activity in a lower area of the temporal lobe known as the *fusiform gyrus*. In addition to such regional-scale involvement, even single neurons have been shown to respond selectively to particular faces, as discussed earlier. Humans and other primates are especially adept at recognizing familiar faces; even though any one face can display a range of expressions. When we see the image of a face on our computer screen, we may, at first, be unsure if the face is a real person or a computer-generated image. However, our observation of a talking face over several seconds is likely to expose most imposters. The movie industry has expended substantial efforts to create computer-animated images that appear fully human. Someday we may even be able enjoy new films starring famous "actors" who are long dead, but are indistinguishable from the real actors. However, because our brains are so expert at distinguishing genuine human from nonhuman images, this technology has proven to be quite challenging.

During face perception, neural networks apparently employ links of some kind to allow the recall of associated memories. The several stages of face processing include recognition of the face, recall of memories associated with the face, and name recall. Face perception of strangers, and especially loved ones, often evokes emotional and other kinds of responses involving widespread brain regions. Selective brain damage can cause all kinds of strange symptoms; for example, earlier we cited Capgras syndrome, a disorder involving delusions that close family members have been replaced by identical-looking imposters. This delusion commonly occurs in patients diagnosed with paranoid schizophrenia, but it can also occur in patients suffering from brain injury and dementia. In essence, this malady is the creation of false models of other persons. Face recognition is also evident in many animals, even across species; for example, crows have been shown to recognize individual human faces.

How are visual images processed and stored in the brain? When we see an unfamiliar image, how do we categorize that image and place it in the proper context of past experiences? Just to cite one example—there are more than three hundred breeds of dog; they range in weight from about four pounds to over two hundred fifty pounds. Yet, when we see one of these dogs, our brains signal "dog" even for breeds that we have never seen before. We also recognize cartoons, drawings, and even abstract paintings representing dogs. Even a very young child knows the difference between a dog and a cat. How do brains accomplish such complex functions? The answers are mostly unknown; however, in the following section, I propose a simple, brain-friendly model—one that appears to be consistent with genuine visual processing.

8.6 CONSTRUCTING AN ANDROID COMPOSED OF NESTED NETWORKS

Several interesting questions are raised by visual processing. One issue concerns nested hierarchy, both in the assumed underlying networks and in the vast number of nested image categories—dog faces within general faces, within general images, and so forth. Another issue concerns the vast number of networks that, in theory, must constitute the bases for image recognition and processing. We address these questions in the context of the simple Christmas-lights model, but with a network twist. We do not attempt to summarize known features of human visual perception networks, which are, at best, understood only in broad terms. Rather than attempting to model perception in an actual human brain, we imagine the inner workings of some artificial intelligence; let's say an android with networks that may share common ground with genuine human networks.

My simplistic approach to visual processing is aimed at a general audience, and it is unlike the narrative that would be expected in a more-focused scientific text. Many scientists have spent their careers studying visual-system details in humans and lower animals; this work could easily fill several large books.[5] One especially important visual process is *feature detection*, in which the nervous system filters complex natural stimuli in order to extract behaviorally relevant information. For example, some neurons in the visual cortex respond strongly to sharp edges in their visual fields; such edges are likely to occur in objects or animals in the environment. By contrast, our fanciful android model attempts to explain perception of more-complex images by "callously" skipping ahead of the line leading to genuine brain science. To proceed otherwise might risk becoming bogged down in a myriad of the "devil's details," the real stuff that is so necessary for the production of good science.

Artificial neural networks consist of interconnected nodes, partly akin to vast networks of brain neurons—the artificial nodes exchange information through links. The links have numeric weights, like weighted graphs, that are adjusted based on the network's experience, making neural nets adaptive to inputs and capable of learning. Like other machine learning tools, artificial neural networks have been used to solve a wide range of tasks that are hard to solve using ordinary rule-based programming, including computer vision and speech recognition. In our example, we assume that our android's artificial neural network has "learned" through trial and error to recognize visual images in a manner similar to current methods of artificial intelligence. However, our fanciful android's abilities are advanced well beyond today's technology.

Let's start discussion of our network model with the simple system composed of the four nodes shown in figure 8-3. For purposes of this discussion, I will expand our earlier Christmas-tree analogue to four trees representing the four lobes of cortical tissue. Thus each node represents a different Christmas tree in your living room, analogous to the occipital parietal, temporal, and frontal lobes of the cerebral cortex. Call this a *tree-scale network*, analogous to a *lobe-scale brain network*, irrespective of whether or not the links between nodes are in place. Just to keep things simple, consider the number of light patterns produced when just two of the four trees produce on-lights, ignoring the other possibilities of three on-lights, four on-lights, or no on-lights. Different network patterns are formed whenever any two nodes are connected. The solid lines indicate networks 1-2, 1-3, and 1-4. The three additional possible two-node networks are 2-3, 2-4, and 3-4, represented by the dashed lines. Thus, four nodes allow six possible networks, assuming all have exactly two active nodes (lights on).

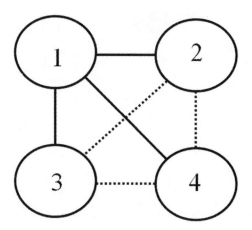

Fig. 8-3.

In our Christmas-lights analogue, each of the tree-scale (lobe-scale) nodes represents a cluster of smaller lights. Let's look inside one of the lighted trees for sub-networks analogous to the *regional scale* of the cerebral cortex, as defined in chapter 3 (i.e., the Brodmann-area scale, see table

3-1). Figure 8-4 shows an example of a ten-node regional-scale system supporting many possible network patterns. In order to simplify our discussion, we characterize each network only by its pattern of on-lights, ignoring the actual links between nodes. As further simplification we consider only networks with exactly half of their nodes active (lights on); let's call these networks "half sub-networks." If exactly five nodes are employed in the ten-node system, 252 possible active network patterns are possible.[6]

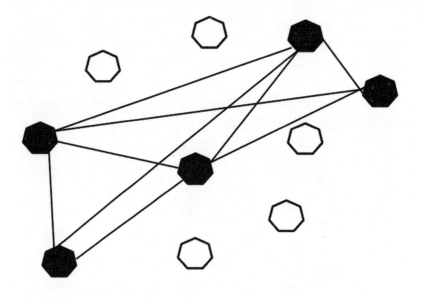

Fig. 8-4.

As indicated in chapter 3 (table 3-1), the cortical macrocolumn scale is about one percent of the regional scale. Thus, when we look closely into any one of the regional-scale active nodes (or light clusters in our Christmas tree) shown in figure 8-4, we find a sub-sub-network (or sub-module) consisting of one hundred nodes. In a one-hundred-node system, there are about 10^{29} possible half sub-networks, each containing fifty active nodes. For a true sense of the size of this number, let's try to put it in perspective: this number of half sub-networks, 10^{29}, is roughly equal to the number of large grains of sand that could be packed inside the earth's volume. Our simplified conceptual network picture is based only on binary states; that is, the lights (nodes) are either on or off. Thus, the links between nodes in this simple picture are assumed to do nothing more than determine which nodes are active (or lighted).

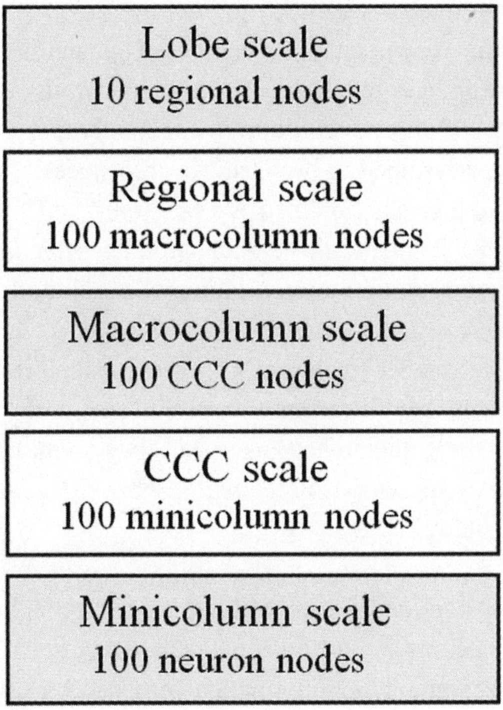

Lobe scale
10 regional nodes

Regional scale
100 macrocolumn nodes

Macrocolumn scale
100 CCC nodes

CCC scale
100 minicolumn nodes

Minicolumn scale
100 neuron nodes

Fig. 8-5.

Given these basic ideas about nested networks, we can now question how visual images of various kinds might be stored, retrieved, and acted upon by our android's networks operating at multiple scales. In short, we employ our android to shed some light on possible consciousness correlates of human recognition of visual images. Consider a thought experiment in which we correlate node (light) patterns in our android with visual image processing. As indicated in figure 8-5, our android brain consists of cortical lobes (lobe-scale nodes), with each lobe containing a ten sub-node (regional-scale) cluster. Each regional-scale cluster is in turn composed of one hundred macrocolumn-scale sub-sub-nodes. This multiscale picture may be continued to progressively smaller scales of nodes within nodes within nodes, as in the case of neurons within minicolumns within cortico-cortical columns (CCC) within macrocolumns. This "fractal-friendly" progression of scale crossing might continue to single neurons, molecules, and perhaps even smaller scales.

In our thought experiment, we imagine that our android recognizes and processes visual images by employing nested networks. After early processing that may or may not conform to genuine brain processes, general visual images are assumed to be separated from faces at intermediate or large scales. If an image is classified as a face, the next level of neural processing occurs at a smaller-scale network, as indicated in figure 8-6, which determines whether the face image represents a real face. The next question, answered at a still-smaller scale, is whether the face is human. And so on down to some smaller (perhaps CCC or minicolumn) scale where the android's "grandmother" is recognized. In this case, many single neurons may respond to the grandmother image, but they might consist of only a very small fraction of neurons in the region or column and be difficult to locate experimentally.

This fanciful android network model addresses several genuine issues that must be faced in the study of real human brains, including the quantitative question of just how many networks are required to recognize and process the vast amounts of information provided by our environment. Since we will later encounter some big numbers in other contexts, I have listed some examples of large numbers in table 8-1 so that we can enjoy an intuitive feeling for the magnitudes involved. OK, maybe I got just a little carried away here; you might consider this table as a "nerd sidebar," but bear with me. The number of grains of sand that could be packed into the volume of the known universe depends on grain size, but my estimate of 10^{91} is a reasonable number. The popular label for the number 10^{100} is the *googol* (not "Google"); its origin is credited to the nine-year-old nephew of an American mathematician. A googol is a mere billion times larger than the "universal sand number."

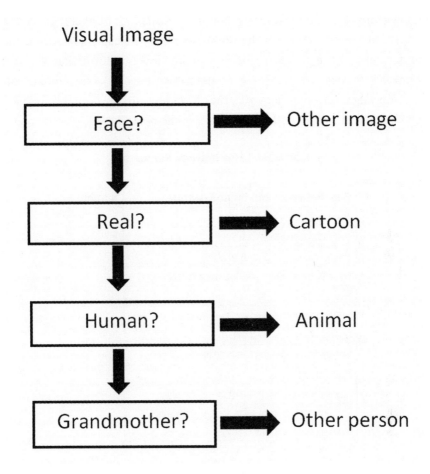

Fig. 8-6.

To imagine the representation of even larger numbers, we can think in terms of the space required to print the numbers. For example, a googol is 1, followed by one hundred zeros; it may be printed in a short paragraph. The number of possible half sub-networks in a ten-billion-cell network of the human brain is roughly $10^{3 \text{ billion}}$. Printing this number in books, consisting of 1 followed by three billion zeros, would require an entire library. Finally, we may detach ourselves from such childish associations with piddling numbers and consider a "real man's number," the *googolplex*, defined as 10^{googol}. Printing out this number would require books filling the known universe. But what's the point of this discussion? Other than having a little fun, it suggests that the brain has plenty of capacity to form the nested networks required to process vast amounts of information obtained from the physical world. However, describing how links turn on and off to form many different networks that form and dissolve quickly is quite another matter, which will be addressed in the next section.

Table 8–1. Large Network Numbers.

Sub–Networks and Other Systems	Number
Half sub-networks in a 4-cell network	6
Half sub-networks in a 10-cell network	252
Macrocolumns in a human brain	10^4
Cortical neurons in a human brain (10 billion)	10^{10}
Half sub-networks in a 100-cell network	10^{29}
Grains of sand packed into Earth's volume	10^{33}
Grains of sand packed into the known universe	10^{91}
Googol (a paragraph of printed zeros)	10^{100}
Half sub-networks in a 1,000-cell network	10^{299}
Half sub-networks in a 10,000-cell network	$10^{3,008}$
Half sub-networks in a million-cell network (one full book of printed zeros)	$10^{301,026}$
Half sub-networks in a 10-billion-cell network (one full library of printed zeros)	$10^{3 \text{ billion}}$
Googolplex (universe full of books with zeros)	10^{googol}

8.7 BRAIN RESONANCE

This section addresses the basic question of how multiple networks and their associated rhythms might interact with each other to orchestrate different mental functions, the "beautiful music of consciousness." In order to prepare for some light technical discussion of brain resonance, I will once again recruit the handy sociological metaphor introduced in chapter 1, consisting of social networks of various sizes operating within a large-scale culture.

We can easily picture the essential circular causality operating in such social systems, the top-down and bottom-up interactions across spatial scales. As a consequence, human societies produce an uncountable number of dynamic patterns when carrying out their daily activities. I have also emphasized the balance between functional segregation and functional integration, a balance evidently required in brains that are both conscious and healthy. In short, different brain regions do different things; they are segregated. At the same time, they cooperate to yield a unified behavior and unified consciousness, so they are integrated. As discussed earlier, the question of how this can be accomplished in human brains is known as the *binding problem.*

In this section, I employ the same sociological metaphor to address possible neural processes underlying brain binding. With this goal in mind, consider several small social groups (networks) based on some common interest of the members. Perhaps each network involves persons who play tennis together; or maybe they share a particular political or religious viewpoint. The networks are, of course, overlapping; each person is probably a member of multiple networks based on different shared interests. The networks are typically *semi-autonomous*—meaning that the members of each network are much more strongly linked to each other than to persons outside that particular network. Otherwise, each network would not exist as a distinct entity. In the example of a tennis club, perhaps nearly all tennis matches take place between several hundred members of the club. However, despite this semi-autonomy, some interactions between separate networks may also occur, for example, if two tennis clubs arrange to send their best players to face competition with another club. Such interactions are often generated at the local level.

Another kind of interaction between social networks can occur as a result of top-down social, political, religious, or other kinds of influences. Imagine several Democratic-based networks and Republican-based networks in the United States that normally interact with each other only weakly or perhaps not at all. Suppose each party nominates a divisive candidate who is disliked by some members of his or her own party. This top-down influence may result in substantial increases in various kinds of novel interactions between the networks that are normally semi-autonomous.

Perhaps some citizens from opposing parties actually get together on many issues; we might label these connections "resonate relationships." As is often said, "politics makes strange bedfellows." I should probably add this disclaimer—any similarity in this story to persons living or dead is purely coincidental (or not).

Here we focus on a very different kind of resonance that may bind neural networks from the top down in a manner that is metaphorically similar to my fanciful sociological resonance. Scientists have developed many mathematical and computer models of brain networks, which may (or may not) represent important features of genuine neural networks. In many, if not most, cases these model networks tend to generate their own natural frequencies, as well as to respond selectively to different input frequencies from external systems. In other words, the model networks can act like antennae that are mainly sensitive to certain input signals, although their actual responses to narrow frequency ranges (or bands) of inputs vary widely. In general, the term *resonance* indicates the tendency for systems to respond selectively to inputs in narrow frequency bands. Such systems typically generate the same natural frequencies spontaneously in response to noise inputs. The labels "natural frequency" and "resonant frequency" mean almost the same thing, but the former is employed more often for spontaneous rhythms, while the latter is normally used to indicate selective responses to external input.

Cell-phone and radio antennae operate on resonance principles, as do animal visual and auditory systems. Every cell phone has a private resonant frequency allowing it to select from thousands of simultaneous calls coded in the electromagnetic fields impacting every phone. TV and radio tuners are based on the same principle, but they allow control of local resonant frequencies by turning the dial to the frequency band matching the desired broadcast station. In a similar vein, mechanical resonance of the basilar membrane in the inner ear enables us to distinguish different audio frequencies in voices and other sounds. It has even been suggested that our sense of smell may involve resonance interactions between odor molecules and vibrations of chemical bonds in our olfactory receptors.

Physical structures like violins, pianos, bridges, power lines, and airplane wings display resonant responses at their natural frequencies. Some resonance is good, as in flute playing; some is bad, as when earthquakes cause buildings to fall. Quartz watches use mechanical resonance to keep time; MRI is based on resonant absorption of electromagnetic energy by protons; atomic energy levels depend on a kind of quantum wavefunction resonance. Molecules have multiple natural vibration frequencies; thus various gases in our atmosphere selectively absorb light or other electromagnetic fields at their resonant frequencies. Resonance is just about everywhere, and as argued here, it is also expected in brain networks that form at multiple scales.

Figure 8-7 pictures two interacting systems X_1 and X_2; these might be neural networks or nearly any kind of systems producing rhythms with their own set of natural frequencies when isolated. For example, the lower system (X_2) might be one of the strings of a violin, and the upper system (X_1) might represent the violin body. In this case, the string produces a set of natural frequencies—the fundamental (f_{21}) and its overtones (f_{22}, f_{23}, f_{24}, and so forth). In our notation the first subscript indicates the network number, and the second subscript indicates the fundamental frequency (1) and the overtones (2, 3, 4, and so on). In the violin string example, the overtones are harmonics, but in most systems the overtones are not harmonics; that is, overtone frequencies are not usually simple multiples of the fundamental. The entire collection of fundamental and overtone frequencies for system X_2 is here assigned the boldface (vector) symbol f_2. Thus, f_2 represents the set of fundamental and overtone frequencies ($f_{21}, f_{22}, f_{23}, f_{24}$ and so forth). Similarly, the collection of natural frequencies of system X_1, perhaps a violin body, is indicated by the symbol f_1.

The horizontal arrows in figure 8-7 indicate any kind of external (perhaps environmental) influences on the two networks coming from some separate system X_0, and the vertical arrows represent direct interactions between the networks X_1 and X_2. These two small-scale networks may be embedded in a global environment that includes thousands of other networks, analogous to two small social networks embedded in a culture. The net (top-down) result of all these external influences (X_0) is represented by the horizontal arrows; such input may itself consist of a broad range of rhythms, including white or colored noise.

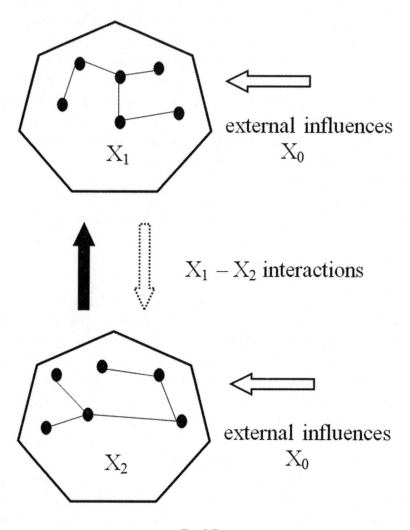

external influences
X_0

X_1

$X_1 - X_2$ interactions

external influences
X_0

X_2

Fig. 8-7.

OK, you ask, just how does figure 8-7 apply to the brain's actual networks? While very few answers are known, we can paint a plausible picture based partly on mathematical models of a broad range of networks found in physical and other systems. In other words, the dynamic patterns produced by many different kinds of systems are governed by differential equations, and these mathematical methods have been verified by numerous experiments. Many plausible models of neural networks are also based on similar methods. Although the actual neural details are not well understood, we can provide some general ideas that are at least partly independent of such missing information. Here we are mostly interested in the binding of distinct networks by top-down resonance to form temporary functional connections; the technical details are left to the notes.

We start with the case when no interactions occur between the two smaller systems X_1 and X_2; that is, the two vertical arrows are removed from figure 8-7. When the top-down (environmental or other) influences, X_0, indicated by the horizontal arrows, are small and random (e.g., white-noise input), the small-scale systems X_1 and X_2 tend to produce rhythms at their own natural frequencies. When this top-down input includes frequencies that match one of the smaller-scale natural frequencies, a relatively large resonant response is expected. The systems X_1 and X_2 might represent mechanical structures like bridges or airplane wings, which can be "driven" (forced into motion) by wind or earthquakes. Alternately, X_1 and X_2 might be tiny cell-phone antennae that select narrow frequency bands of electromagnetic waves from the broad spectrum of cell-tower broadcasts represented by X_0. For our purposes, X_1 and X_2 represent small- or intermediate-scale neural networks embedded in a large-scale network, or perhaps a global field of synaptic action, represented by X_0, which acts top-down on the smaller networks.

In the next case, we remove the environmental influences (horizontal arrows) and continue removal of the downward interaction (dashed arrow) of the upper system (X_1) on the lower system (X_2) so that only the upward interaction remains (black arrow). System X_2 will then tend to force its own activity on system X_1 at some combination of X_2's natural frequencies. In more-technical terms, X_2 is said to "drive" X_1. When only a single natural frequency is produced by system X_2, it may be labeled a "pacemaker" of the activity in system X_1. Similarly, an artificial heart pacemaker (X_2) is a small, battery-operated device that senses when the heart (X_1) is beating irregularly or too slowly. X_2 sends a signal that makes the heart beat at the correct pace. For a long time, many neuroscientists believed that the human alpha rhythm in the cerebral cortex was caused by a pacemaker in the thalamus, an idea that is now largely discredited.

Now let's include the feedback interaction of system X_1 back on system X_2 to our picture; this feedback is represented by the dashed arrow. If this interaction is weak and the external influences (X_0, horizontal arrows) are also small, we may reasonably approximate system X_2 as semi-autonomous, that is, relatively isolated from other parts of the brain to first approximation. However, if the interactions represented by the two vertical arrows are both moderate to strong compared to the internal network links, a new combined system $X_1 + X_2$ is created. The natural frequencies of the new system will generally bear no simple relationship to the natural frequencies of the isolated systems. The expectation of such feedback from the cortex to the thalamus is just one reason why the alpha-pacemaker claim has largely disappeared from scientific papers. On the other hand, local alpha rhythms generated in thalamo-cortical networks ($X_1 + X_2$) are believed to occur regularly, along with more-global alpha rhythms (X_0) perhaps generated exclusively in the cortex. All of these rhythms, generated at multiple scales, may provide genuine signatures of consciousness; however, finding the locations and spatial extent of the brain tissue generating each of these rhythms is a notoriously difficult task.

8.8 CROSS-FREQUENCY COUPLING AND BINDING BY RESONANCE

In this section, I suggest a means by which pairs of small-scale neural networks (X_1 and X_2) can be bound together (functionally connected) by some third system (X_0). The system X_0 might be a network of a similar scale, perhaps located in the thalamus, or it might be a large-scale system acting top-down on the smaller-scale networks. In either case, this conceptual framework, based on network resonance, directly addresses the fundamental binding problem of brain science: How do distinct networks cooperate to produce a unified behavior and consciousness, while maintaining sufficient autonomy to perform essential sensory, motor, and other specific tasks? If our fanciful black cloud, introduced in chapter 6, can accomplish this, maybe brains do similar things. While some may be disdainful of this analogue, I argue in this section that similar kinds of resonant behaviors might be anticipated by well-established neuroscience, albeit aided by some essential mathematics. Not to worry, while the ideas are outlined in the running text, the mathematics is confined to the notes.

Cross-frequency coupling is the creation of functional connections of some kind between rhythms of different frequencies. For example, the phases of theta and gamma oscillations in different brain areas can line up in certain circumstances, forming a kind of functional connection called *phase coupling*. Here we are not concerned with the actual nature of such functional coupling; rather, we consider the more-general issue of resonance phenomena as a means of producing network coupling of some kind. I argue that such coupling allows separate networks to interact selectively even when their natural frequencies don't match. This idea, binding by resonance, opens the door to many new phenomena for neuroscientists to consider in their quest to understand more about consciousness.

To explain how binding by resonance might work, we again consider two systems, X_1 and X_2, that are physically connected in some way, as represented by the vertical arrows in figure 8-7. Our goal here is to apply basic ideas to a broad range of possible brain networks. Thus, we focus on the special case of weak physical connections between X_1 and X_2 and question how the external influence X_0 (horizontal arrows) is likely to affect the functional connection between X_1 and X_2. That is, how might the frequency content of the external influences modulate the strength of functional connection between X_1 and X_2?

To address this fundamental question, two mathematicians interested in neuroscience, Eugene Izhikevich and Frank Hoppensteadt, employed a very general network model based on the mathematics of nonlinear differential equations.[7] By focusing first on mathematical rather than on physical or other issues, some very general results were obtained. The general mathematical model might represent many kinds of physical or biological systems, including neural networks formed at different spatial scales. For our purposes, the strength and appeal of this abstract mathematical approach is that such models, although far from perfect, provide an overall picture of the general dynamic behavior to be expected in many kinds of actual systems. The main goal of such models is to be *useful*; strong correspondence to genuine brain tissue is neither expected nor required. By "useful" I mean that such models facilitate better experimental designs in projects aimed to confirm or deny various postulates about brain functions. As outlined in the endnotes,[8] the essential results of these studies are as follows:

- If the physical connections between X_1 and X_2 are strong, new natural frequencies emerge in the combined system $X_1 + X_2$. In this case, the individual networks lose their distinct identities; they cannot be treated as semi-autonomous. For example, suppose that the single system $X_1 + X_2$ represents the cortical tissue of the primary visual cortex (V_1). If this system is strongly interconnected (within itself) as we expect, any imagined separation into subsystems X_1 and X_2 is expected to substantially degrade or perhaps completely obliterate normal visual processing. On the other hand, if network X_1 represents the primary visual cortex and X_2 represents some other area of cortical tissue that is only weakly connected physically to X_1, perhaps the functional connection between X_1 and X_2 can turn off or on quickly, as outlined below.

- If networks X_1 and X_2 are only weakly connected physically, relatively strong functional connections (binding by resonance) between X_1 and X_2 can still be produced by the external influence of a third system, X_0, provided that special combinations of natural frequencies occurring in the three systems (X_1, X_2, X_0) satisfy certain resonance relations. For example, the processing of visual information involving interactions between networks X_1 and X_2 might be modulated by some global field or network X_0, a system that is mostly determined by other brain operations reacting to the external world.

The most obvious case of possible binding by resonance occurs when systems X_1 and X_2 have the same sets or overlapping sets of natural frequencies. One such example occurs when the two fundamental frequencies match, that is, when f_{11} and f_{21} are equal, providing one way to achieve binding by synchrony, a special case of binding by resonance. Another case is when one local system's fundamental frequency matches an overtone of the other system, as when f_{11} equals f_{22}.

The above discussion suggests that binding by resonance allows rhythmic systems with only weak physical connections to exhibit strong functional coupling due to "matching" natural frequencies. As suggested by the quotation marks, the natural frequencies *need not be equal* to *be matching*. Rather, modern studies show that strong "binding" (functional connections) can occur in systems that are only weakly connected physically, provided the right combinations of natural frequencies are matching. We leave the mathematical details to the endnotes;[9] however, one interesting case of resonant binding is predicted when one of the natural frequencies of system X_0 equals the *difference* between the natural frequencies of systems X_1 and X_2.

We can tie the above example specifically to the brain rhythms closely associated with both general consciousness and specific mental states. Suppose that system X_0 is a separate network generating a 5 Hz theta rhythm. Assume also that many smaller-scale networks with gamma-band natural frequencies are embedded in the brain. The 5 Hz theta rhythm is then predicted by this simple model to strengthen the functional connection between two gamma networks with natural frequencies of 37 and 42 Hz. In another example, a large-scale 10 Hz cortical alpha rhythm is predicted to strengthen the functional connection between two gamma networks with 36 and 46 Hz natural frequencies, as indicated in figure 8-8. This same large-scale alpha rhythm would also be predicted to strengthen the functional connections between two small-scale (thalamo-cortical) 10 Hz alpha rhythms. The precise frequency numbers cited here are not meant to be taken seriously, and, in any case, many more examples of resonance binding are predicted by the model as outlined in the endnotes.[10] The main point is this: *Interaction strengths between pairs of distinct (semi-autonomous) local networks, embedded in a third system, are expected to be modulated by the third system because of resonant interactions among the three systems.* This general idea does not rest on the exclusive presence of simple (narrow-band) rhythms—in real brains, we should expect much more complicated resonant interactions than those provided by these simple examples.

Local networks acting together

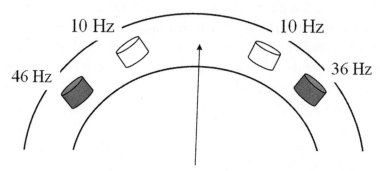

10 Hz global resonance in a spherical shell

Fig. 8-8.

This mathematical model suggests important roles for cross-frequency interactions in general and binding by resonance in particular in the consciousness challenge. For example, experimental studies in humans and animals have demonstrated that slow oscillations, typically in the alpha or theta bands, can modulate the fast gamma-band rhythms associated with local networks and cognition. For example, in studies associated with memory storage, human EEG recordings reveal a relationship between gamma amplitude and the phase of theta oscillations. In another example, modulation of local theta rhythms by the large-scale delta rhythms of deep sleep has been reported, also possibly facilitating long-term memory storage.

A number of additional experimental studies suggest that cross-frequency coupling of different rhythms may serve a number of functional roles in the brain, including computation, language, learning, and visual perception.[11] In particular, coupling strength differs across brain areas according to task, and it can change rapidly in response to sensory, motor, and mental events. Such rhythmic coupling is correlated with mental performance. While high-frequency brain activity tends to reflect local regions of cortical processing, low-frequency brain rhythms are dynamically coupled across distributed brain regions. Thus rhythmic coupling (binding by resonance) may provide an efficient means to transfer information from slow-acting, large-scale brain networks to the fast, local cortical processing, thus integrating functional systems across multiple spatial-temporal scales. The potential implications of this picture for the brain's many overlapping conscious, pre-conscious, and unconscious systems is intriguing, to say the least. I offer this claim because many of the unanswered questions about the sources of mental activity are closely tied to the formation of functional network connections that turn or off quickly.

8.9 SUMMARY

As in numerous applications in physics and engineering, models of brain systems may be based on either networks or continuous media. Continuum models assume that the media (substance) fill the space they occupy. This "mass-action approach," known formally in a broad class of physical systems as *continuum mechanics*, ignores the fact that matter is made of networks of molecules, which themselves consist of networks of atoms. Thus, matter is not really continuous; however, when treating problems involving length scales that are much greater than intermolecular distances, such models are often quite accurate. Brains, like other complex systems, can be modeled as either networks or continuous media; the two views are often compatible. Such models are employed to illuminate various kinds of mental activity as well as the more-general brain binding problem, the means by which individual networks act together to create a uniform consciousness.

The human connectome is the map of neural connections within the brain, essentially its wiring diagram. Brain links may be defined at several or perhaps many levels (spatial scales). If single neurons are represented as network nodes, we might call this microlevel a *level 0 network*. At the same time, larger brain structures like cortical regions or columns may serve as larger-scale networks, labeled *level 1*, *level 2*, and so forth up to the *global* or entire-brain scale. Graph theory provides a mathematical basis for network study. The labels *graph*, *vertex*, and *edge* correspond to the scientific terms *network*, *node*, and *link*, respectively.

Graphs or networks may be characterized in many ways, for example, by the average path length, clustering features, and various measures of multiscale modular structure, that is, networks within networks within networks, and so forth. A small-world network is a type of graph or network typically consisting of local clusters with dense interconnecting links, along with sparse links between separate clusters. The influence of non-local interactions is often an important feature of small-world networks. The application of graph theory to brain anatomy involves many messy, non-mathematical issues. A basic question concerns spatial scale: Which anatomical structures should serve as the appropriate nodes?

Cortical connectivity at large scales is dominated by white matter axons, especially the cortico-cortical axons. While connectome knowledge is one important aspect of brain science, many reject the idea that connectome knowledge implies an "understanding" of the brain. One confounding issue involves weighted links between nodes, accounting for different interaction strengths. Myelin, synaptic, or other kinds of plasticity may be critical for healthy mental states, essentially providing plastic weights to network links. In this context, note that a broad range of psychiatric and other disorders has recently been associated with myelin defects in the white matter. Myelin determines the speed of action potential propagation. Hence, white matter diseases are linked to the disruption in the timing of arrival of action potentials in local networks.

In order to demonstrate several network-related ideas in the context of brain science, we consider how the human visual system might carry out complex tasks. *Visual perception* is the conscious experience of processing visual information; in particular, face perception is critically important for most human social interactions. During face perception, neural networks must employ links of some kind to allow the recall of associated memories. The several stages of face processing include face recognition, recall of associated memories, and name recall.

A fictitious android is proposed to model visual perception. In the proposed four lobe-scale brain networks, each lobe-scale node contains ten regional-scale nodes; each regional-scale node contains one hundred macrocolumn-scale nodes, and so on down to the single-neuron scale. Our fanciful android system is analogous to four Christmas trees representing the four brain lobes. Each tree has light clusters within light clusters, representing nodes at different scales. Given this simple model of how visual images might be stored, retrieved, and acted upon by our android, we consider a thought experiment in which node (light) patterns encode visual-image processing.

Our android recognizes and processes visual images by employing its nested networks. General visual images are first separated from faces at large scales. If an image is classified as a face, the next level of neural processing occurs in a smaller-scale network that determines whether the face image represents a real face or some other image, like a cartoon face. The next question, answered at a still-smaller scale, is whether the face is human. And so on down to some even-smaller scale where the android's "grandmother" is recognized. This fanciful android network addresses several genuine issues that must be faced in the study of real human brains, including the quantitative question of just how many networks are required to recognize and process the vast information sets provided by our environment.

An essential question of brain science is how multiple networks and their associated rhythms interact with each other to orchestrate different mental functions. Suppose two small-scale neural networks (X_1 and X_2) are embedded in a large-scale network or other brain environment, X_0. If the networks X_1 and X_2 are only weakly linked by physical means, strong functional connections (binding) between X_1 and X_2 can still occur due to the external influences of network X_0. This coupling is predicted, provided combinations of natural frequencies occurring in the three systems (X_1, X_2, X_0) satisfy certain resonance relations. For example, if each of the three systems (X_1, X_2, X_0) has a single (fundamental) natural frequency (f_{11}, f_{21}, f_{01}) when isolated, strong functional coupling between X_1 and X_2 is expected when $f_{01} = f_{11} - f_{21}$. Thus, a 5 Hz theta rhythm can modulate the strength of functional connection between two gamma networks with natural frequencies of 38 and 43 Hz, or two other gamma networks with natural frequencies of 35 and 40 Hz. In short, cross-frequency interactions provide theoretical and experimental mechanisms for top-down or bottom-up interactions between brain networks operating at different spatial scales.

What does this general picture of multiscale resonance in brains have to do with the emergence of consciousness? In introducing the hard problem in the next chapter, I will discuss the *multiscale conjecture*, which claims that the emergence of consciousness requires production of dynamic patterns at multiple scales. While many assume the existence of some special "C-scale" that encodes consciousness, the multiscale conjecture posits that no such C-scale is required. In contrast, many different scales are required to operate in concert for consciousness to emerge. The multiscale conjecture can be interpreted in two ways—as a stand-alone materialistic model of consciousness or, alternately, as an introduction to more-dualistic interpretations.

Chapter 9

INTRODUCTION TO THE HARD PROBLEM

9.1 THE CHICKEN-SOUP TEST

Discussions in the first eight chapters mostly addressed the so-called easy problem; that is, scientific studies of the so-called consciousness correlates or signatures of consciousness. By contrast, chapters 9 and 10 are more focused on the hard problem, the origin of consciousness itself. Of necessity, this new material must contain much more in the way of speculation than the first eight chapters; however, speculations will be responsibly limited, remaining consistent with known physical laws and established neuroscience. We shall see, however, that such modest restrictions leave us quite free to explore a wide range of strange and even weird intellectual territory. We begin with a brief summary of some mainstream scientific findings and then tiptoe cautiously into more-speculative domains.

Consciousness is consistently associated with certain kinds of brain complexity, notably the nested hierarchy of tissue in the cerebral cortex with its multiscale columnar structure, plus the abundance of non-local white matter connections. Recall the significance of the classical "non-local" label: First, this feature allows any pair of brain locations to exchange signals without influencing the intervening tissue. Second, such long-range cortical anatomy facilitates small-world phenomena in which sparsely connected networks may operate more freely along the spectrum of more localized activity to globally coherent activity, thereby substantially magnifying network complexity. The relationship between the dynamics of complex system and small-world links, whether they are fixed structural links or other kinds of connections, is observed or at least implied in a large number of physical, economic, social, biological, and other systems. Disease spread, Internet links, stock markets, and wars provide just some of the prominent examples.

When pondering the enormous complexity of conscious beings or even nonconscious life, it will prove fruitful to contrast such "mega-complexity" with certain physical systems that display more moderate levels of complexity. By adopting this approach we aim to expand our insight into the similarities and differences appearing along the spectrum of progressively more-complex systems. One interesting class of physical systems that snugly fits this intellectual strategy is fluid turbulence (note that "fluid" refers not just to liquids)—common examples include the air turbulence experienced in airline travel, smoke rising from cigarettes, blood flow through defective heart valves, tornados, and pots of boiling soup. Our interest in fluid turbulence stems from its inherent multiscale character, a feature largely responsible for its complexity.

Consider your grandmother's chicken soup. As it sits atop the stove, the heated soup in the pot may rise to the top in fluid structures called *convection cells*; the soup then moves to the sides and descends back to the bottom, forming alternating patterns of vertical fluid flow, essentially "soup rhythms" with oscillation frequencies that depend on both the soup and soup pot's properties. *Convection cell rhythms* develop only under certain environmental conditions—depending on the temperature and consistency of the soup, the size and shape of the soup pot, and so forth. Thus, convection cells constitute dynamic patterns that serve as *signatures of convection*; they indicate the state of the fluid in a manner analogous to the brain's various *signatures of consciousness*.

Convection cells also occur in the atmosphere, in the ocean above underwater volcanoes, and in the sun. Unlike the mostly unknown origins of dynamic brain patterns associated with consciousness, basic macroscopic equations governing fluid turbulence are known. Turbulence models employ complicated mathematics; however, computer solutions of dynamic behaviors are mostly limited to statistical predictions. This point is not a criticism of meteorology; statistical predictions save lives. Rather my central point is this—even with the fundamental mathematics of fluid flow securely in our toolbox, we are severely limited in our computer predictions of dynamic patterns created in turbulent fluids. This discussion of turbulence is meant to remind us of the very real limitations on the computer processing required to produce future advances in artificial intelligence.

Why are our fluid-flow predictions so limited? What makes turbulence complex? Turbulence is composed of swirling fluid structures, *vortices* and *eddies*, that reveal themselves over a broad range of spatial scales— little swirls within larger swirls, nested hierarchies. Turbulent complexity is characterized by several major features—the first is the multiscale distribution of its dynamic patterns and their attendant rhythms. Much of the kinetic energy of turbulent motion is contained in large-scale fluid structures; this energy and its attendant information cascades (top-down) from these large-scale structures to smaller scales. This process continues, creating smaller and smaller structures that produce a nested hierarchy of eddies and vortices down to molecular scales where the basic fluid equations become invalid. Thus, cross-scale interactions, the familiar circular causality characteristic of most complex systems, occurs in turbulence as it apparently occurs in the cerebral cortex. A major difference, however, is that unlike fluid systems, the cerebral cortex contains an extensive multiscale *anatomical structural hierarchy*, namely cortical columns that can facilitate multiscale dynamic patterns of information. Furthermore, the cerebral cortex is interconnected by a massive system of non-local or small-world axons forming most of the white matter layer; such long-distance connections are believed to contribute substantially to brain complexity, just as they do in many physical, social and other systems.

The second manner in which turbulence displays complexity is through its exquisite sensitivity to environmental conditions; turbulent flow is inherently chaotic. Suppose a severe weather system moves across Kansas. Over the past several decades, meteorologists have identified certain dynamic patterns of wind speed, pressure, temperature, and so forth that substantially raise the chances of local tornado activity. However, such features cannot predict precisely where and when each tornado will occur; nor can they accurately predict tornado size and intensity. The major reason for these limitations is that chaotic systems are exquisitely sensitive to both *initial* and *boundary conditions*. The initial conditions indicate the state of weather at some time before tornado production. The *boundary conditions* refer to the environmental conditions in which the weather evolves. Change these conditions by just a little bit, let's say by one thousandth of one percent, and the macroscopic weather system might change substantially.

Only relatively recently did chaos gain widespread scientific appreciation. Chaos is characterized by hypersensitivity to initial and boundary conditions because microscopic fluid or molecular patterns act (bottom-up) to contribute strong influences on macroscopic behavior. Essentially, microscopic information "infects" macroscopic properties. Between the 1970s and the early 1990s it became progressively more apparent to most physical scientists that even seemingly simple systems can produce chaotic patterns of behavior. Examples include weights bouncing on simple springs or balls circulating in roulette wheels.

Chaos is a purely classical phenomenon; no quantum effects are required. *Thus, classical physics, with no boost from quantum mechanics, has forever banished the rigid determinism of nineteenth-century physics to the dustbin of history.* This relatively new finding has important implications for philosophy and brain science, including several issues surrounding the ever-present free-will controversy. The remnants of "autistic determinism," as I like to call it, may now survive mainly in the unconscious brain modules of a few out-of-date scientists and philosophers. The realization of the profound significance of chaos led to the famous question posed by a pioneering meteorologist: Does the flap of a butterfly's wings in Brazil set off a tornado in Texas?[1]

Given this recent history, whenever I hear of some new exaggerated claim from artificial intelligence, I'm tempted to ask the following, "Has your computer even mastered 'simple' turbulence? That is, have you managed to pass any kind of *butterfly* or *chicken-soup test*?" Such an imagined test would access the computer's ability to deal with genuine complexity. Models of turbulence can be very useful—however, because of chaos and the critical influences of cross-scale interactions, truly accurate computer simulations of turbulent systems like boiling soup may be forever beyond human technology. Life and especially consciousness appear to depend on systems that are far more complex than fluid turbulence, yet claims that true artificial consciousness is just around the corner are commonplace. *Does this really make sense?*

Chaos impacts nearly all complex systems, and it apparently plays a major role in the development of living systems. We have repeatedly discussed the creation of larger-scale systems formed from smaller-scale systems—brains are made of cells composed of macromolecules composed of atoms, and so forth. The larger-scale systems develop novel features of their own, unanticipated larger-scale properties that are entirely absent from the smaller-scale systems. These emergent properties allow the large-scale systems to act top-down on the smaller-scale systems, which can then act bottom-up on larger-scale systems, closing cycles of circular causality. But many have questioned how larger-scale systems can be anything more than just the sum of their parts. One clear answer is that larger-scale systems are formed under the influence of their environments (the boundary conditions); such system development is subject to the whims of chaos.

To cite just one example of the expected profound impact of chaos on both living and nonliving systems, imagine a billion "earths" evolving in environments that differ only slightly from one earth to another over the entire five billion years of their existence. Our imagined near-equal earth environments include similar suns, initial conditions at their formations, meteor impacts, and so forth. Chaos science suggests that the resulting ecosystems, if they were to exist at all, would probably differ widely from one to another. Only one earth might produce dinosaurs and mammals; other earths might produce unimagined creatures, but perhaps most earths would produce no life at all. In this view the development of the first life on Earth (*abiogenesis*) was a lucky accident; but once life appeared, natural selection allowed life as we see it to evolve. A possible counterargument to the random interpretation of abiogenesis is that some unknown organizing principle, perhaps related in some manner to entropy or information, facilitated the generation of life from molecular systems.

It seems that the fundamental constants of our universe fall within a very narrow range compatible with life; our universe may be *fine-tuned* for life. Questions about life's existence generally fall under the rubric of the so-called anthropic principle. The *weak anthropic principle* just states the obvious— only a universe capable of eventually supporting life will contain living beings that ask such questions. This interpretation often employs the imagined *multiverse*, a population of universes from which life's selection-bias originates. In short, we are here—so it is unremarkable that our particular universe is fine-tuned for life; otherwise we wouldn't be here to talk about it. Stronger versions of the anthropic principle offer a broader range of possible implications of the apparent fine-tuning of our universe that allowed life to develop.[2]

9.2 WHAT IS LIFE?

Turbulent systems are complex, much more complex that most systems normally studied in the engineering and physical sciences; however, turbulent complexity pales in comparison to *life*. But since most life apparently lacks consciousness, questions about life's features are appropriate for discussion in the context of the hard problem. If nonconscious life is complex, how does it differ from conscious life, which is apparently even more complex? Even relatively "simple" life includes many of the most complex systems known to science. By most measures, living systems are far more complex than any of today's most advanced computers or other artificial systems.

One can argue plausibly that life is also more complex than any artificial system likely to be created in the near future. Living systems are composed of cells; each cell contains perhaps ten billion interacting molecules, with many molecules composed of thousands or even hundreds of thousands of atoms. In the course of dynamic interactions between molecules, each cell seems to act somewhat like a natural supercomputer, an information-processing and replicating system of enormous complexity. While simple cell models can be very useful for limited purposes, accurate simulation of genuine cell dynamics appears to be well beyond the ability of current science. One reason for this limitation is that chemical bonds are governed by the principles of quantum mechanics, and accurate quantum calculations (based on Schrödinger's equation) are currently possible only for small collections of interacting particles. While I fully acknowledge that computational progress in physics is impressive and ongoing, the implied computational gap between small particle systems and the macromolecules of life is truly enormous. Even just a single living cell is far more complex than any current computer.

The category "life" covers a continuum of progressively more complex nested hierarchies of systems and subsystems as summarized by the oversimplified list in table 9-1. The complexity of various living systems is expressed here in terms of different "levels of life," ranging from Level 0 (nonliving) to Level I (cells) to the highest known level, Level IV, here associated with multiple members of a species. Table 9-1 represents only one of several plausible ways to represent this nested hierarchy of living systems. We could, for example, have expanded the list to include ecosystems consisting of both living and nonliving parts of an environment. The highest level of known biological organization is the *biosphere*, the collection of all of Earth's ecosystems; however, table 9-1 appears adequate for our current purposes.

Table 9-1. Levels of Living Systems.

Life Level	Entity	Example Function	Example Structure
0	Macromolecules	Store information	DNA
0	Organelles	Transfer energy	Molecular aggregates, mitochondria
I	Cells	Semi-isolate molecular systems within membrane boundary	Eukaryotes
II	Organ systems	Pump blood and process sensory input	Cell populations: cardiovascular and nervous systems
III	Organisms	Obtain energy from the environment and reproduce	Trees, bacteria, ants, snakes
IV	Populations	Multiple members of a species	Forest, ant colony, human city

Life is neither a substance nor a "thing" in the usual sense; rather, it is a complex process expected to exhibit many or all the following traits. Life is composed of "societies" of one or more cells that self-regulate their internal environments, the process known as *homeostasis*. Transformation of chemical energy is employed to maintain internal molecular organization, the process of *metabolism*. Living organisms adapt; they possess the ability to change in response to the external environment; they also synthesize proteins using genetic instructions encoded by DNA. Living organisms typically grow in size, respond to external stimuli, and reproduce themselves. They take energy from the external environment—directly from the sun, from plants or animal tissue, or even from underwater hot gas sources (hydrothermal vents). Living systems use this energy to self-organize their internal environments. In this manner, life processes create a decrease in internal entropy, the so-called inaccessible information to be discussed further in chapter 10.

Given this realistic picture of bona fide life, provocative and exaggerated labels like *artificial life* or *artificial intelligence* should perhaps be acknowledged more explicitly as catchy tags designed to promote favored science or engineering projects rather than representing science accurately. My negative attitude toward this overblown language is not meant to downplay the tremendous advances made in so-called artificial life and artificial intelligence. Rather, my point is that such fancy and exaggerated vocabulary can easily mislead the general public, as well as some scientists, who may be seduced by the provocative promotional pronouncements. I don't object to the tag "artificial intelligence," as long as "intelligence" is not equated with "consciousness."

As far as we know, consciousness first requires life, but far more is required. That is, all known life provides necessary but far from sufficient conditions for consciousness to occur. Life began on Earth about four billion years ago, roughly five hundred million years or so after the Earth was formed. The origin of Earth's life is unknown, although a number of plausible hypotheses have been advanced. Most scientists believe that life began with some sort of self-organizing molecules. The various forms of life include plants, animals, fungi, and bacteria. Viruses present an ambiguous case; they are *almost-living organisms*. Viruses possess genes, evolve by natural selection, and make copies of themselves; however, they require a host cell to carry out these functions. As discussed earlier, consciousness apparently involves interactions between semi-conscious subsystems (modules). By analogy, viruses may be categorized as "semi-life."

Life involves a vast number of subsystems organized in nested hierarchies. Given that even "simple" life is so incredibly complex, we might want to revisit the question of conscious computers. We should question if it makes sense to suppose that any computer, one that is far simpler than even a single living cell, could attain even rudimentary consciousness. For instance, I am currently looking out through the window to my backyard. I'm pretty sure that my pine tree is not conscious; however, my tree seems to possess a better shot at gaining rudimentary consciousness than any computer system that falls far short of the tree's inherit complexity. Forget the Turing test; let the imagined computer take some sort of chicken-soup test; that is, demonstrate a level of complexity similar to "simple" life, which operates far above "simple" turbulence.

9.3 THE NEUROLOGY OF CONSCIOUSNESS

The title of this section mirrors that of Steven Laureys and Giulio Tononi's (LT) compilation *The Neurology of Consciousness*, which provides a comprehensive and authoritative picture of consciousness correlates from the perspective of mainstream neuroscience.[3] LT is a sequel and companion to an earlier collection of scientific articles. It shows greater editorial control, offering a narrower range of subjects treated in a more-organized fashion; it also omits some of the more-controversial chapters in the original collection. The following list covers several of the main issues presented in LT, plus nuanced interpretations based on our earlier discussions. The bullet points below represent a summary of information gained from multiple chapters of LT plus several other sources cited in my earlier chapters.

- A central message from neurology is that proper functioning of cortical-thalamic systems is essential for consciousness to occur. By contrast, inactivation or damage to many other brain areas need not abolish consciousness.

- The critical importance of cortical-thalamic systems to consciousness presents a stark contrast to the role of the cerebellum, a brain structure containing more neurons than the cerebral cortex. The cerebellum is also richly endowed with both internal connections and connections to other brain structures; it often shows selective activation during mental tasks and may respond to emotional events. Yet, in marked contrast to cortical-thalamic systems, widespread damage to the cerebellum produces only minimal effects on consciousness.

- Loss of consciousness during seizures, slow-wave sleep, anesthesia, and vegetative states often involves similar brain areas, but this picture does not reveal a simple one-to-one correspondence. For example, excessive synchronization during seizures results in loss of consciousness, whereas moderate synchronization (of special kinds) is apparently required for consciousness to occur.

- Consciousness may be distinctly disassociated with other brain functions like responses to sensory inputs, motor control, attention, language, memory, and even the sense of self. Consciousness does not require language or sense of self and does not simply reduce to just attention or memory.

- When the left hemisphere is isolated, either permanently with split-brain surgery or temporarily with drugs, it acts like a fully functioning brain with a full sense of self. An isolated left hemisphere has no concern or even knowledge of his or her missing former partner, the right hemisphere.

- The case of the isolated right hemisphere is a bit more complicated, mainly because language is usually produced only by the left hemisphere. In the past, some scientists denied consciousness to the right hemisphere. But the preponderance of recent scientific studies indicates that the right hemisphere operates with private sensory and motor channels and has its own memories and likes and dislikes. The right hemisphere enjoys a genuine, if somewhat-impoverished, sense of self, a self at least as robust as those observed in other primates lacking language.

- Consciousness is not an object occupying a fixed location or instant in time; rather, it is an ongoing process distributed over both time and space within the brain. The dynamic patterns of consciousness require several hundred milliseconds to develop. Various dynamic patterns, measured at multiple spatial and temporal scales, provide reliable signatures of conscious. Brain rhythms are an essential feature of these dynamic patterns.
- While simple models like the Christmas-lights brain can sometimes be useful, consciousness is not simple; rather, it emerges from interactions within and between the brain's complex subsystems. Complexity is essential to consciousness, although no single measure of "complexity" is widely agreed upon. Healthy brains operate between the extremes of functional isolation and global coherence, states associated with high dynamic complexity. Various disease states are associated with the extreme ends of this broad spectrum of dynamic behaviors.
- Brains contain many interacting unconscious, semi-conscious, preconscious, and conscious subsystems; brains are somewhat like social systems in this respect. Unconscious subsystems (modules) do most of the information processing of input from the external environment, and much of this information is stored in memory. Such subliminal information influences our actions, often through intuition; however, only small parts of such information ever reach conscious awareness.

The list above outlines, in very general terms, a useful overview of existing solutions to the easy problem; it offers an essential scientific framework supporting our speculative investigations of the hard problem outlined in chapters 9 and 10. We will continue this quest by looking into the range of consciousness levels exhibited in different living systems.

9.4 MULTIPLE LEVELS OF CONSCIOUSNESS

Table 9–2. Levels of Consciousness.

Consciousness Level	Example Species	Characteristic Behavior	Representative Brain Structure
0	Plant life	Response to weather	None
I	Reptiles	Moves through space	Brainstem
II	Mammals	Social interactions	Limbic system
III	Humans	Sense of time	Cerebral cortex
IV	Large-scale intelligence	Advanced knowledge	Interacting brains

The label "consciousness" does not refer to a single brain state; rather, the distinctions between unconscious, semi-conscious, pre-conscious, and conscious conditions are many and varied. Examples of graded consciousness levels include sleep, which consists of different stages of semi-conscious states. Different consciousness levels are also demonstrated by patients in progressive stages of Alzheimer's disease. Consciousness increases over time in developing young children. We allow at least some level of consciousness in lower mammals, and so forth. Multiple levels of consciousness are suggested by the simple model outlined in table 9-2.[4] In this view, all life consists of complex systems; however, plant life does not seem to achieve even minimal consciousness. On the other hand, snakes and other reptiles possess rudimentary awareness of their surroundings, here identified as Level I consciousness. Such awareness allows reptiles to move through their spatial environment to find food, react to threats, and so forth. Level II mammals do all of this, but they also interact with their peers, providing a number of benefits including protection from predators and reproduction. Level II consciousness apparently operates mainly in the "here and now," that is, with little or no planning for the future.

Level III human consciousness is associated with a strong sense of self in relation to the past, present, and future. That is, creatures operating at Level III can create mental models of the future that aid in making current decisions. Level IV consciousness is speculative; however, it seems to be a natural extension of the first four levels, as discussed in the next section. Table 9-2 represents just one of several ways to quantify levels of consciousness. One might, for example, plausibility break down each major level into sublevels. For example, rats might be rated Level II.1; smart chimps, Level II.6; patients with moderate Alzheimer's disease, Level III.0; and so forth. Any such attempt to quantify consciousness levels is bound to represent a vast oversimplification, but table 9-2 seems like as good a start as any.

9.5 LARGE-SCALE INTELLIGENCE?

Is large-scale consciousness, representing the Level IV category of table 9-2, theoretically possible? In chapter 2, we discussed the fundamental problem of identity. In particular, consciousness involves dynamic patterns created through the interactions between brain subsystems. We suggested that each of us can remain an individual only by limiting communication with other persons. In evolutionary terms, the development of language may have substantially increased the human consciousness levels; for purposes of discussion, let's say something like an expansion from Level III.0 to Level III.5. One may guess that such a difference in consciousness level is similar to that which occurs between isolated left and right hemispheres. I am suggesting that language acts to increase consciousness level, both through interacting with other persons as well as facilitating communication between the semi-isolated subsystems of individual brains. In this limited sense, the extension of consciousness levels in modern thinkers, let's say, up to something like Level IV.0 may not be especially controversial.

Let me now pose a crazy-sounding question—could a larger-scale consciousness emerge from a group of strongly interacting individuals of the same tribe? Perhaps mobs, soldiers in combat, encounter groups, and even some marriages might provide examples of rudimentary steps toward larger-scale or *group consciousness*. The idea may seem ridiculous to some, but suppose we imagine some progressive process where more and more information barriers between individuals are purged so that the group becomes more and more functionally integrated. Is there any level of functional integration where the group might be considered a single consciousness?

Does the fact that participants remain physically separated make a difference? Suppose we facilitate functional integration by progressively adding more physical connections between brains using an imagined advanced technology that selectively relays action potentials between brains. If a group consciousness were in fact created in this manner, what would happen to individual consciousness? Could group and individual consciousness coexist? If so, could each become aware of the other's existence? Would individual consciousness become subservient to the group, a sort of mob psychology? How similar are these questions to the issues raised in regard to interactions between the multiple semi-conscious systems in single brains?

If this thought experiment seems a bit far-fetched, I suggest an unequivocal example from real life. *Conjoined twins* ("Siamese twins") are identical twins formed in the uterus either by an incomplete splitting of the fertilized egg or a partial rejoining of an egg that had split earlier. This occurs roughly once in every 200,000 births; about 25 percent survive. My first example is the famed "Scottish brothers" of the fifteenth century, essentially two heads sharing the same body. They had separate brains, so we naturally grant them separate consciousness. But, they carried the same genes and were never physically separated, sharing the same environment all their lives. Can we perhaps infer some overlap in their consciousness? Were there ever periods when one could say they had a single consciousness? After all, most people generally consider the normal right-left brain partnership to be a single individual—how similar was the Scottish brothers' partnership?

My next example is twins joined at the head with partial brain sharing. Do they have one consciousness or two? It seems that the adjective "partial" is critical here. Lori and George Schappell, Americans born in 1961, are joined at their foreheads and share 30 percent of their brain tissue. George—who was born female but identifies as male—has performed widely as a country singer and has won music awards. During George's performances, Lori remains quiet and attempts to become "invisible." From these reports, Lori and George appear to each have their own entirely separate consciousness. But imagine several pairs of twins with progressively more brain matter in common; at what point do two consciousnesses become one? How similar are these questions about conjoined twins to issues concerning split-brain patients or interactions between the semi-conscious subsystems in single intact brains?

These thoughts involving strongly interacting social groups, conjoined twins, and cross-hemispheric brain connections may raise doubts that an individual's consciousness must be confined entirely to his own brain. Suppose, for example, that we study an experimental volunteer named Joe by chemically isolating his brain hemispheres over some short time interval. Later, Joe's isolated left hemisphere (Joe-left) is reconnected to his right hemisphere (Joe-right). By all accounts, the articulate Joe-left never notices any difference between his disconnected and connected states. However, special laboratory tests would easily reveal important differences between these two states. Should we not grant the paired hemispheres named "Joe" (Joe-right-plus-left) a somewhat-higher level of consciousness than the isolated Joe-left?

In my view, one implication of Joe's experience is that individuals operating in social groups can attain somewhat-higher levels of consciousness than humans who are substantially isolated from their peers. I don't deny the shipwrecked Robinson Crusoe human level consciousness but rather suggest that his consciousness is somewhat impoverished by his isolation; no fully functioning man is an island. This picture again supports the influence of various means of interpersonal communication on higher levels of consciousness, by means of voice, text, mathematics, art, music, tactile exchange (hugs), and perhaps even sex. OK, maybe I got a little carried away here; I can envision a future headline from the *National Enquirer*, "Scientist Claims Hanky-Panky in Bed Extends Consciousness!"

9.6 MULTISCALE PATTERNS OF INFORMATION

We have seen that experimental signatures of consciousness are observed at multiple levels of organization (spatial scales). Two competing interpretations of these data come to mind. First, perhaps consciousness is encoded in dynamic patterns at some special, but as yet undetermined, spatial scale. In chapter 8, we labeled this the *C-scale*. In this view, consciousness signatures observed at other scales are mere by-products of C-scale dynamic behavior. Perhaps, for example, the special C-scale is that of the single neuron; that is, consciousness is encoded in patterns at the single-neuron level, a view seemingly adopted by many neuroscientists and artificial-intelligence scientists. This view, which is closely related to our metaphorical Christmas-lights brain, implies that an artificial brain consisting of some hundred billion or so "artificial neurons," if appropriately interconnected, might achieve consciousness. The units representing artificial neurons in this case might obey simple input-output rules, or they might behave in a more-complicated manner. But, in either case, top-down influences on the elementary units would be secondary to the actual production of consciousness. Neuroscience, in this common view, takes on a strong reductionist flavor—the single-neuron C-scale, the level where consciousness "resides," is relatively small-scale. C-scale dynamic patterns are then imagined to create (bottom-up) the larger-scale patterns that scientists record with various experimental techniques. Circular causality plays no major role in this C-scale picture.

An alternate interpretation to the C-scale proposal is that consciousness is fundamentally a multiscale phenomenon; I call this the *multiscale conjecture*. In this view, consciousness is encoded by the dynamic patterns occurring at multiple scales; consciousness is intimately associated with cross-scale interactions, both bottom-up and top-down. Circular causality is an essential feature of this picture. In the following discussions, I favor the multiscale conjecture based mainly on several overlapping arguments as follows:

- I know of no compelling evidence for the existence of some unique C-scale; consciousness signatures are observed over a range of spatial scales. In addition to the single-neuron scale, plausible arguments favoring the minicolumn scale (one hundred neurons) have been advanced, for example. In chapter 10 we will look more closely at sub-neuron scales (molecular and below).

- Possible consciousness signatures at a number of intermediate scales plus molecular and even smaller scales are largely inaccessible to today's science. This raises the plausible possibility that many more signatures of consciousness will be discovered in the future. In particular, the apparent relationship of long-term memory to changes in protein molecules in synapses is not well understood.

- Consciousness seems to occur only in systems with exceptionally high complexity. Multiscale dynamic behavior and associated pattern formation is a typical feature of known complex systems. Fluid turbulence and the worldwide human social system provide just two of many examples of multiscale complexity from disparate scientific fields.

- Unlike the cerebellum, the cerebral cortex is essential for the occurrence of consciousness. Each of these two brain structures contains more than ten billion neurons; however, only the cerebral cortex possesses a rich, multiscale columnar structure. In addition, the white matter layer consists largely of massive interconnections of ten billion non-local (small-world) cortico-cortical axons. Both of these anatomical features can be expected to substantially facilitate system complexity.

The multiscale conjecture implies that the so-called connectome, while unquestionably important, embodies only a small part of the mystery of consciousness. To emphasize this point, imagine that aliens visit Earth and attempt to understand the human social system. They are quick to grasp essential features of our "social connectome," including interactions between individuals, cities, and nations by means of television, the web, and so forth. But, even with perfect knowledge of the social connectome, the aliens would still be a long way from understanding wars, economic conditions, religious differences, politics, and numerous other social phenomena caused mostly by multi-scale interactions. Furthermore, our society can be profoundly influenced by the interactions across timescales as facilitated by libraries, history teachers, movies, and so forth. With this analogy in mind, it appears to me that some pundits have oversold the importance of the brain connectome by suggesting that its full knowledge implies something close to an "understanding" of consciousness.

9.7 WHAT IS INFORMATION?

Scientific questions about the consciousness challenge are intimately connected to the general idea of information. We are deeply interested in relationships between the inner world of consciousness and the physical world of brains. Effective discussion of potential links between such seemingly disparate phenomena can be facilitated by adopting language that overlaps both worlds. With this goal in mind, labels like *information, memory, signals, network, knowledge,* and *calculation* are routinely applied in both the brain and computer sciences. These common labels often indicate useful metaphorical relationships between biological and physical processes having important functional features in common, but they may also differ fundamentally at the core. In this context Raymond Tallis, a medical scientist and philosopher of mind, warns of the pitfalls of such overlapping language. Arguments based on terms like *information*, which have so many disguised interpretations, can fool our conscious and unconscious systems, hiding logical flaws that fail to stand up in linguistic daylight. As Tallis expresses it, "the deployment of such terms conjures away the barriers between man and machine."[5]

We have consistently linked dynamic patterns to various states of consciousness. Such patterns carry information; thus for our purposes the label "dynamic pattern" may be considered identical to "dynamic pattern of information" since patterns must, by definition, carry information of some sort. But just what is our definition of "information"? It has several common meanings depending on context.[6] In everyday language, "information" typically involves processing, manipulating, and organizing data in a way that adds to the knowledge of the person receiving it. For our purposes, however, the label "information" refers to *scientific information*, which differs from its everyday use, mainly because scientific information can be expressed in numbers that measure its "size." Furthermore, scientific information is divorced from *decoding* or *meaning*, features typically supplied by the sender and receiver. A man asks his girlfriend, "Will you marry me?" Her yes or no answer consists of only a single bit of information, but it has substantial meaning to both of them. Lovers and scientists think differently about "information."

The scientific information content of a cell phone text, baseball, computer disk, movie, or nearly any entity is measured in bits.[7] For example, suppose you flip a coin ten times and communicate the sequence (pattern) of heads and tails to someone else. Let heads be listed as one and tails be listed as zero; each coin flip then produces one bit of information, and the sequence of ten coin flips contains ten bits of information. The fact that this information is probably of no interest to the receiver is irrelevant to the actual, well-defined quantity of information. In a more-practical example, ASCII (American Standard Code for Information Interchange) codes each letter or symbol on your keyboard in a sequence of seven bits, for example the letter "e" is coded as 1100101 (101 in base-10 arithmetic). A message may then be sent in the form of a series of seven-bit blocks, each block representing a single symbol, with the individual blocks separated by the space code (0100000, or 32 in base-10 arithmetic). In this case, all meaning in the information is supplied by the sender and receiver; *meaning does not influence the quantity of information in the sequence* or *pattern*. Earthquakes, propaganda, music, art, traffic noise, expressions of emotion, and strangers smiling in restaurants all convey information, whether or not the events are observed by scientific instruments or living systems. The so-called receivers may be trees, rocks, humans, stars, or just about anything.

Consider physical objects like baseballs in the context of scientific information. A baseball consists of a cork center wound with layers of string and covered with a stitched leather covering; baseballs possess certain large-scale physical and informational properties. An approximate macroscopic description of a moving baseball might consist of its size, shape, mass, location, and velocity. More-detailed information is needed to accurately predict its trajectory, however, including its spin about three axes and the condition of its surface. Spin provides the pitcher with an effective weapon against the batter, the curveball. Before 1920, the spitball was also legal; earlier pitchers tried many ways to alter baseball surfaces, scuffing it with foreign objects, applying spit, and even tobacco juice, thereby adding to trajectory uncertainty, both real and imagined, by batters. The macroscopic state of a moving baseball might be described in moderate detail, including the surface coordinates of its raised stitches contributing to its trajectory, by perhaps fifty variables, each with, let's say, three-digit precision (10 bits) or roughly 500 bits of information.

If matter behaved classically and was continuous down to arbitrarily small scales, a physical object like a baseball would contain an infinite amount of information.[8] The amount of information in any *finite* description would then depend on the precision assigned to the locations and velocities of each small piece (e.g., an electron) of the baseball. By contrast, in quantum theory, material objects are believed to contain finite, rather than infinite, information because of the fundamental limitations on precision dictated by the famous *uncertainty principle*. While many nonphysicists may think of this principle as just some practical limitation on experiments with elementary particles, it has far more profound implications in the general context of information, as discussed below and again in chapter 10.

A complete microscopic description of a baseball in terms of the locations and velocities of its elementary particles would consist of something like 10^{25} bits of information, or about ten million times the number of grains of sand in all the beaches and deserts on Earth. Thus, all macroscopic objects carry enormous amounts of information in the form of microscopic dynamic patterns. Most of this information at the microscopic scale is inaccessible to ordinary observation; such *inaccessible information* is identified as *entropy* in chapter 10. While the microscopic information is normally inaccessible, this partial knowledge barrier is not the most fundamental known to science; rather a kind of "deep entropy" or "hidden reality" is introduced below and discussed in more depth in chapter 10.

According to the uncertainty principle of quantum mechanics, it is fundamentally impossible to gain more than a certain amount of information about any physical object. With macroscopic objects, this limitation is mostly unnoticed, but at the scale of electrons and other elementary particles, the uncertainty principle rules. One implication of this fundamental principle is that it is impossible to add precision or information beyond 10^{25} bits to this description of a baseball. If any additional baseball information actually exists, it is *unknowable* according to current science. This label is meant to be taken literally: "Unknowable" means absolutely and forever unknowable (in contrast, "inaccessible" means not easily known).

Entropy is information contained in a physical system that is inaccessible, that is, normally invisible to us. Early in the twentieth century, scientists discovered the existence of a much deeper and more fundamental knowledge barrier, a *hidden reality*[9] whose existence is now widely accepted (at least by physicists), but whose actual nature remains the subject of ongoing controversy. This hidden reality has been associated with a number of fancy labels, including *wavefunction collapse*, *many worlds*, *multiverse*, *implicate order*, *coherent histories*, and more. Discussion of these competing interpretations of quantum mechanics is beyond the scope of this book, although I will outline some of the issues in the endnotes and in chapter 10. But my main point is this—although the true nature of this hidden reality is currently unknown (and possibly forever unknowable), there seems to be little doubt of its influence—not only on our everyday lives but also on the very existence of our universe.

The three underlying principles of physics—the basic laws that govern our universe, are general relativity, quantum mechanics, and thermodynamics. In chapter 10 we will see that all three laws provide fundamental information barriers: relativity limits speed; quantum mechanics limits quantity; and thermodynamics limits quality of information transfer. For example, the fact that the microscopic information contained in material objects is finite rather than infinite (as expected in classical physics) seems to embody profound scientific significance. We have claimed that consciousness is intimately associated with patterns of information; in fact we apparently remain individuals only by limiting information transfer with other persons. Perhaps this implied connection between physics and consciousness is only a coincidence. However, I contend that any serious study of the hard problem should follow this path to modern physics and see where it leads. With this philosophical position in hand, chapter 10 will associate the hidden reality of physics with an extension of the usual idea of scientific information; I call it *ultra-information* for lack of a better term. I will speculate that this hidden reality might have some connection to consciousness. This approach may be labeled the *RQTC conjecture*, standing for relativity, quantum mechanics, thermodynamics, and consciousness.

9.8 SUMMARY

This chapter introduces my head-on approach to confront the hard problem of consciousness based on three background bodies of information: (1) scientific data on the easy problem—the nature of consciousness signatures covered in the first eight chapters; (2) more-general observations from several disparate branches of science; and (3) plausible speculations restrained by known scientific principles. Our conceptual framework first envisions a broad spectrum of progressively more-complex systems ranging from physical systems like turbulent fluids to non-conscious life to conscious beings.

Examples of turbulence include cigarette smoke, blood flow, tornados, and boiling soup. Turbulent complexity is revealed by the multiscale distribution of its dynamic patterns and by its exquisite sensitivity to environmental conditions; turbulent flow is inherently chaotic. As a consequence of chaos, computer predictions of the dynamic patterns created in turbulent fluids like weather systems and boiling soup are severely limited; they are necessarily statistical forecasts. Appreciation of the critical importance of chaos to science and philosophy deepened substantially after the pioneering studies of meteorologist Edward Lorenz. His famous question posed in 1972, "Does the flap of a butterfly's wings in Brazil set off a tornado in Texas?" led to a flood of studies of the so-called butterfly effect in many systems previously thought to behave simply. This work demolished the rigid determinism of nineteenth-century science, even for purely classical (non-quantum) systems.

While turbulent systems are unquestionably complex, such complexity pales in comparison to life. Most life apparently lacks consciousness, but consideration of life's properties provides an entry point to study the consciousness challenge. If non-conscious life is complex, how does it differ from conscious life, which seems to be much more complex? The complexity of various living systems may be expressed as different "life levels," ranging from Level 0 (nonliving) to Level I (cells) to the highest known level, Level IV, which is associated with multiple members of a species.

A comprehensive picture of consciousness correlates from the perspective of mainstream neuroscience is provided by *The Neurology of Consciousness*. One of its main messages is that proper functioning of cortical-thalamic systems is essential to consciousness. By contrast, widespread damage to the cerebellum and many other brain structures has minimal effects on consciousness.

Consciousness is not a single brain state; rather, distinctions between unconscious, semi-conscious, pre-conscious, and conscious conditions are many and varied. Examples of graded consciousness levels include sleep, which consists of different stages of semi-conscious states. Unconscious subsystems (modules) do most of the information processing of sensory input from the external environment, and much of this information is stored in memory. Healthy brains operate between the extremes of functional isolation and global coherence, states associated with high dynamic complexity. Similarly to the quantification of life levels, we quantify consciousness levels as follows, with examples in parentheses: Level 0 (plants), Level I (reptiles), Level II (mammals), Level III (humans), and Level IV (putative larger-scale consciousness).

Consciousness involves dynamic patterns created through the interactions between brain subsystems. If such patterns of information extend beyond individual brains, perhaps the beginnings of larger-scale consciousness may develop in groups of strongly interacting individuals. Mobs, soldiers in combat, encounter groups, and some marriages provide possible examples of rudimentary steps toward larger-scale or group consciousness. One might speculate that an intact brain may possess a somewhat-higher consciousness level than an isolated left hemisphere. Or an individual strongly embedded in a social system may attain a somewhat-higher consciousness level than a socially isolated person. This general idea suggests implications for more-extreme kinds of group behavior like mass hysteria or cults. Could such groups ever be credited with some kind of large-scale consciousness? This idea seems strange, but maybe no more so than computer consciousness.

We have seen that experimental signatures of consciousness are observed at multiple levels of organization. Two competing interpretations of these data are thus: (1) Consciousness is encoded in dynamic patterns at some special spatial scale, the so-called C-scale. In this view, consciousness signatures observed at other scales are mere by-products of C-scale dynamic behavior. (2) Consciousness is fundamentally a multiscale phenomenon; this position is labeled the *multiscale conjecture*. In this latter view, consciousness is encoded by the dynamic patterns occurring at multiple scales; it is intimately associated with cross-scale interactions, both bottom-up and top-down. In this view, circular causality in brains is essential to the emergence of consciousness.

Scientific questions about the consciousness challenge appear to be intimately connected to the general idea of information, which may indicate relationships between the inner world of consciousness and the physical world of brains. Scientific information is measured in bits; such information is fully divorced from meaning, which is typically supplied by senders and receivers. The three fundamental laws of physics, believed to govern the behavior of the entire universe, are informational laws, laws that place fundamental limits on information: relativity limits speed; quantum mechanics limits quantity; and thermodynamics limits quality of information transfer.

Scientists have discovered fundamental knowledge barriers associated with a hidden reality or "ghost world." Although the true nature of this hidden reality is unknown, its influence is felt everywhere—not only on our everyday lives but also on the very existence of our universe. In chapter 10 we will associate the hidden reality with an extension of the usual idea of information, called *ultra-information*, and speculate that this hidden reality might have some connection to consciousness. I call this approach the *RQTC conjecture*, representing relativity, quantum mechanics, thermodynamics, and consciousness.

Chapter 10

MULTISCALE SPECULATIONS ON THE HARD PROBLEM

10.1 BETTING ON LONG SHOTS

Some scientists and philosophers claim that the so-called hard problem is just an illusion. In their view, consciousness somehow follows directly from the brain's massive network interactions, end of story. An often-cited source of this position is *Consciousness Explained*, by the prominent philosopher Daniel Dennett.[1] His critics have enjoyed a good old time with this provocative title, suggesting alternate tags like *Consciousness Denied* and *Consciousness Explained Away*. Such conflicts remind us that intelligent people often come to divergent opinions, even when the same evidence is available to all. My guess is that philosophical positions on the consciousness question stem partly from each person's personal comfort level with unsolved or even unsolvable mysteries. Some fields, most notably mathematics, deal with questions that are unequivocally either right or wrong; if you really hate ignorance, you probably love mathematics. Also, if you hate ignorance, maybe you tend to trivialize mysteries because the problems are just too hard and irritating—at least that's my personal take on the consciousness controversy.

Looking toward the other end of the "mystery and ignorance spectrum," I'm reminded of an astute observation by a colleague that "psychiatrists love things they don't understand." I haven't conducted a survey to check this claim, but given medicine's modest understanding of mental illnesses, it makes sense. Another source of attitudes about mystery and ignorance is the artificial-intelligence pioneer Marvin Minsky—he once stated that he had decided against a physics career because physics problems seemed "profound and solvable."[2] In contrast, Minsky found problems of intelligence to be more appealing because they were insurmountable, in his words "hopelessly profound."

So apparently some of us find good value in pursuing deep scientific questions that are unlikely to be answered in the near future. We must then be satisfied with much more modest goals like shedding just a little additional light on the hidden mysteries. Fortunately for the proper operation of society, we have competent plumbers, truck drivers, engineers, medical doctors, and so forth to deal with the essential day-to-day, practical, and solvable issues. That said, there is no reason why one cannot be a practical person by day, while contemplating scientific and philosophical long shots deep into the night.

Like many budding scientists I chose my doctoral research based on its intellectual challenge without much thought of future employment. My original field was theoretical plasma physics, which, among other things, provides the essential scientific support for *controlled fusion*, a technology in which energy production is based on the same process that causes the sun to shine. This technology requires herculean efforts, from both the physics and the engineering perspectives. An old joke among scientists is that fusion energy is just twenty years away, *and always will be*. Scientists entering this field seem to be betting on a long shot.

Other long-shot candidates include string theory,[3] the search for extraterrestrial intelligence (SETI), parapsychology (psi), and the hard problem of consciousness. I don't mean to imply that these disparate efforts rest on equal footings, but all seem to face substantial barriers in the near or even distant future. String theory is a modern physics framework in which each elementary particle consists of a tiny vibrating string; distinct particles are then determined by the various natural frequencies of the string's oscillatory rhythms. This may sound simple, until you discover that mathematical consistency requires the putative strings to operate in a half dozen or more extra space dimensions beyond the usual three. Judged only on its formidable intellectual challenge, string theory earns very high status. However, to be considered genuine science, even the most sophisticated mathematics must be able to connect with experiments, and experimental verification or falsification of string theory seems to be a long way off.

What about psi research, including extrasensory perception and so forth? Psi has been strongly criticized by many scientists. I am also skeptical of psi, but my opinion is not worth much because I have never studied the field. I do have the impression, however, that some criticism is unfairly based on guilt by association with inferior or deeply flawed psi studies.[4] To put this picture in a broader perspective, note that even good scientists sometimes grow bad apples—nonsense is too often published in many scientific fields. But mainstream science tends to be self-correcting through the process of replication or invalidation of other scientists' works. The experimental designs of these verification efforts require supporting theoretical frameworks to guide scientists on exactly what to look for, plus just how and where to look for it. The main barrier to good psi research seems to stem from its lack of a solid theoretical framework to guide definitive experiments.

The next long-shot candidate on my short list is SETI. The likelihood that advanced civilizations exist in our universe seems high; for me it appears nearly certain.[5] However, I still consider SETI to be a real long shot, partly because of light's universal speed limit and partly because I can't see that humans have much to offer advanced creatures. I'm guessing that such "folks" would rather spend their resources interacting with more-advanced or at least peer civilizations. In a famous episode of *The Twilight Zone*, "friendly" aliens come for a visit. Their apparent benevolence is evidenced by their coded booklet *To Serve Man*. Unfortunately, it is discovered too late that *To Serve Man* is a cookbook. One suspects, however, that advanced civilizations can obtain their protein much easier than by stuffing earthlings into soup pots. This same argument seems to apply to most other things we may think we have to offer; my guess is that in the big picture we're just not that important.

My final long-shot candidate is the hard problem of consciousness. Solving the hard problem may seem like the ultimate impossible challenge; however, we will follow this quest as far as we can, aiming to shed just a little more light on the proverbial "landscape" of mysteries, probabilities, and possibilities.

10.2 MULTISCALE MEMORY AND CONSCIOUSNESS

I have promised to address the hard problem in the context of established science. But in so doing, I will be hedging my bets to some degree. The *multiscale conjecture* is, in itself, consistent with both the materialistic and at least some of the nonmaterialistic positions on consciousness outlined in chapter 2. My main point here is that this general framework, in which *consciousness is based on the dynamic patterns of multiple interacting scales*, appears far closer to the truth than simple models like the Christmas-lights brain. The multiscale conjecture argues forcefully against philosophical positions that appear to trivialize the consciousness challenge. In essence I am claiming that consciousness seems to require systems at least as complex as ordinary life, which consists of interacting multiscale structures. Thus, I am proposing two distinct intellectual domains in which the multiscale conjecture may operate. First, the multiscale conjecture is to be taken seriously as a stand-alone idea, independent of questions about materialism and the hard problem. Second, the multiscale conjecture may provide a tentative bridge connecting brain information patterns to minimally materialistic or perhaps even nonmaterialistic conceptual frameworks underlying the hard problem.

As presented thus far, the multiscale conjecture has focused mostly on cross-scale interactions between the whole brain and various intermediate-scale structures down to the single neuron. But why stop at this cellular scale? If the multiscale conjecture is correct, we are faced with the following fundamental question: Where do we make the small-scale cutoff below which physical structures may be treated simply? Or, said another way: Where does the fractal-like character of the nervous system stop? Whereas many neuroscientists (implicitly or explicitly) have adopted the single-cell scale for this barrier, I see no obvious need for this assumption. While I am not aware of any well-established signatures of conscious at subcellular scales, this data gap may reflect only technological limitations of our measuring devices. Given the technical barriers to possible small-scale signatures, we may instead look into the alternate entity known as *memory*, which is intimately associated with consciousness.

The label "memory" refers to a familiar but, nevertheless, strange collection of phenomena, some closely aligned with awareness. We remember genuine past experiences; thus memory differs both from *imagination* and *perception*, the awareness of current sensory information. Recalling dramatic events like those of September 11, 2001, can elicit vivid visual memories of people and objects located nearby on such occasions. Most of us remember where we were and who we were with on that day. For example, I happened to be evaluating research proposals in the Washington, DC, area, while attending a meeting of a National Institutes of Health (NIH) study section. I can now visualize a picture of the conference table and the other scientists when a young woman interrupted our meeting with the news. She generously offered to reschedule our airline flights, but I quickly realized that my only real travel options were ground-based. As my little story shows, memory can be closely tied to strong emotional experiences. Memory is also an intimate part of our self-awareness, and several fundamental questions about memory come to mind. Where are various kinds of memory stored, and how are they accessed? What are the physical representations of memory, the means of memory information storage called *memory traces* or *engrams*?

Cognitive scientists make relatively sharp distinctions between *short-term* and *long-term memory*, but long-term memory consists of several subcategories including *episodic memory* (of specific past events), *semantic memory* (knowledge about the external world), *procedural memory* (use of objects or body movements), and *emotional memory*. Long-term memory is thought to be widely distributed throughout the brain, but different memory categories seem to be preferentially associated with the regions most involved with the original experiences. The *cerebellum* and *limbic system* (an extensive collection of interconnected subcortical structures) are linked to procedural and emotional memories, respectively. At least some aspects of memory appear to be essential for the presence of genuine consciousness.

Short-term memory is typically associated with dynamic spatial patterns of neural activity and modifications of synaptic strengths. Long-term memory is known to involve construction of new proteins within cells as well as new synaptic pathways reinforcing the strengths of neuron-neuron connections. This general picture, however, provides almost no detail about how vast amounts of memory information, seemingly required for normal human functions, are stored and retrieved. The physical representations of memory, the *engrams*, are largely unknown. Is it plausible that all of the vast information content of the long-term memory of an adult human could be stored in new cell assembly (network) formation, especially given that many of these same cells apparently participate in perception and processing of current events? If so, are different kinds of memory stored in networks simultaneously operating at different spatial scales? Is it possible that important aspects of memory are stored at smaller scales, say in protein molecules or even at still-smaller scales? Let's not forget the incredible complexity of single cells that act like natural supercomputers by processing information and replicating large numbers of molecules. The ratio of the diameter of a neuron cell body to a typical protein scale is about ten thousand; thus the corresponding volume ratio is $(10^5)^3$ or 10^{15}, indicating the maximum number of protein molecules that could fit snugly inside each cell. The actual number may be more like 10^{10}, roughly equal to the number of neurons in the cerebral cortex or the Earth's human population.

Small-scale memory storage associated with awareness may not seem far-fetched if we recall that several other kinds of "memory" are stored at cellular and molecular scales. The success of vaccinations against disease is based on *immunological memory* in which immune systems exposed to pathogens store this contact information in "memory cells" able to mount chemical responses to new exposures. *Antibodies*, which are protein molecules produced in these cells, then identify and neutralize foreign bodies. Immunological memory works like this: An *antigen* is any foreign body that evokes an immune response either by itself or after combining with a larger molecule like a protein. White blood cells identify specific antigens, which then become bound to antigen-specific receptors on the surface of the white blood cells. The combined antigen–white blood cells undergo cell divisions, and the resulting "offspring" differentiate into cells serving multiple functions, including "attack cells." These cells launch attacks against all antigens that initiated the immune response in the first place. In short, immunological memory is stored at the cellular and molecular levels.

Genetic memory in the form of species-specific genes is passed between generations as DNA molecules. DNA contains the information needed to construct specific protein molecules making up the cells of the offspring, essentially providing the blueprint for the construction of all parts of the new living creature. DNA also carries the information determining when and where in the body different genes turn on and off. In summary, short-term, long-term, immunological, and genetic memories all require information storage, but on progressively longer timescales. Short-term memory is associated with dynamic patterns and cell-assembly formation; immunologic and genetic memories are stored at smaller spatial scales including cellular and molecular scales. The usual sharp distinction between short- and long-term memories makes for convenient discussion, but it masks a continuum of processes involving interactions across both time and spatial scales.

10.3 DIGGING DEEP TO EVEN-SMALLER SCALES

Figure 10-1 summarizes cellular and lower levels of the nested hierarchy of matter and life—the Planck scale, elementary particles, atoms, molecules, and cells. The Planck scale is believed to represent the limiting scale of space-time itself, a sort of fundamental voxel (3D pixel) size. That is, space-time is believed to take on a "foamy" structure at the tiniest scales. *Electrons* and *quarks* are fundamental particles; as far as is known, they cannot be broken down into smaller units; *neutrons* and *protons* consist of quarks. All ordinary matter is composed of *elements*, substances consisting of one kind of *atom*. Atoms are composed of protons and neutrons in a central nucleus surrounded by electrons in outer shells. Each macromolecule of life, interacting within cells, may contain hundreds of thousands of atoms. The ratio of the smallest physical scale that can currently be studied (10^{-16} cm) to the Planck length (10^{-33} cm) is 10^{17}, about equal to the ratio of the Earth-sun separation to the size of a living cell. This huge range of scales allows plenty of room for all kinds of sub-electron structures and strange behaviors, but such knowledge is well hidden from current science.

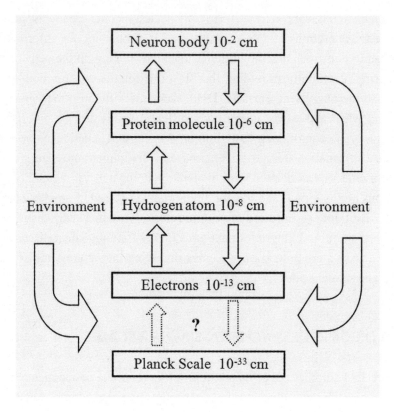

Fig. 10-1.

At each of the higher levels in this nested hierarchy, new structures possessing novel properties emerge from lower scales; the construction of protein or DNA molecules is believed to obey the rules of quantum mechanics at the atomic scale and so forth. Thus, for example, quantum mechanics may be gainfully employed to shed light on specific aspects of a protein's molecular bonds. But, this connection across spatial scales seems to have led some scientists and philosophers to embrace a major fallacy—namely that biochemistry is simply applied physics. While no features of macromolecules are known to violate the principles of quantum mechanics, this consistent agreement does not mean that scientists can *predict* large-scale properties from interactions at the atomic level. Rather, one may argue plausibly that the macromolecules attain novel properties that cannot be predicted from atomic properties, *not even in principle*, for the reasons that follow.

Imagine a "super-duper quantum computer" that simulates the construction of macromolecules like proteins, DNA, and so forth. Let the computer run simulations of interacting atoms subject to different environmental conditions (the boundary and initial conditions), as depicted by the large arrows in figure 10-1. Let the computer run for years while it lists and describes each of the new molecular structures that it computes. At any time we might then count how many molecules our computer has simulated, and try to estimate how this number compares with the total number of *possible molecules*; that is, all imagined molecules that are consistent with the principles of quantum mechanics. This issue depends, of course, on the computer's speed. If we naively assume infinite speed, the issue becomes moot; however, computer speeds are apparently limited by universal laws, as discussed below.

Mathematician and neuroscientist Alwyn Scott has estimated the number of theoretically possible distinct molecules.[6] The number of atoms is 95, excluding the noble gasses that rarely react with other atoms. *Valence* is a measure of the number of chemical bonds that may be formed by each atom. Scott's estimate of the maximum possible number of distinct molecules formed from N atoms is $95^{N/2}$ or roughly 10^N for our purposes. The molecules of living cells, the proteins, DNA, and so forth may contain thousands of atoms, but even if N is limited to a relatively small number of atoms, say one hundred, the number of possible distinct molecules is googol-sized (10^{100}). A mass containing one of each different molecule could just barely be squeezed into the volume of the known universe. By contrast, the number of known distinct molecules is only about 10^7. A mass consisting of one molecule of each kind would fit easily into a tiny corner of a single cell. Notwithstanding the provocative phrases like "end of science" and "theory of everything" that sometimes appear in the science press, we can guess that chemistry will exist as a separate and developing field for as long as there is intelligent life to study it. This little analysis again reminds us that science is accomplished with experiments performed at many different scales, and the nature of cross-scale interactions is often poorly understood. Much of the physics of the twentieth and twenty-first centuries has been focused on the establishment of such cross-scale relationships in various media, superconducting materials providing just one such example. Analogous developments in brain science are now only in their very early stages.[7] These kinds of studies (which I like to tag "crossing the fractal divide") can also be focused on the multiscale brain interactions that may be essential for consciousness to emerge.

Scientists know of nothing in chemistry that is inconsistent with quantum mechanics and have recruited selected aspects of quantum theory to better understand chemical bonds. But the rules governing chemical behavior were developed at the same level (spatial scale) as human observations of these phenomena. We should also not downplay the critical role of the environment (boundary and initial conditions) in any dynamic system, including chemicals in solution. One can say (correctly) that the laws of quantum mechanics are more "fundamental" than those of chemistry, but this may be a rather-hollow sentiment for a practicing chemical engineer. Once a molecular property is known, scientists can often rationalize this property in terms of quantum theory; however, we apparently have no realistic chance of accurately simulating protein molecules bottom-up from fundamental quantum mechanics—certainly not in practice—perhaps not even in principle. Rather, biochemistry relies on the emergent properties of macromolecules; such properties result in no small part from the environmental conditions indicated by the large arrows in figure 10-1. It is even possible that these environmental conditions encompass some sort of fundamental organizational principles, perhaps some currently hidden aspect of information, as discussed below. Once the self-replicating molecules of life form, the well-established principle of Darwin's natural selection kicks in—but how does life come about in the first place? We don't know, but in any case, it's safe to insist that *biochemistry is far more than just applied physics.*

Some supporters of strong reductionism imply that computer technology will eventually solve the enormous barriers of crossing hierarchical levels from the bottom-up. One way to address this issue is to estimate fundamental limits on future computer technology. Computer speeds are measured in *floating point operations per second* (FLOPS). A hand calculator may operate at 10 FLOPS; your desktop computer may operate a billion times faster at 10^{10} FLOPS. Today's fastest supercomputers operate near 10^{15} FLOPS. The physics literature contains several ideas about fundamental limits on computer speed, including esoteric quantum computation using black holes. With such arguments, quantum physicist Seth Lloyd has suggested a limit of 10^{105} FLOPS, which he has presented as a fundamental barrier based on an imagined ultimate computer composed of all the energy and matter in the known universe.[8] If we imagine this "computer-verse" computing continuously from the origin of the big bang to the present day, the computer will have performed "only"10^{122} operations.

Much has been written about future computer advances, including quantum computers. However, Lloyd's ideas suggest to me that crossing even one major level in the hierarchy of matter using "bottom-up" brute force computer calculations can probably never be accomplished *in practice*, maybe not even *in principle*. Note that we are interested here in "principles" that govern natural processes observed by humans, not magic created by supernatural beings. This limitation does not, however, prevent science from establishing properties at some macroscopic level (based on experiments at the same level) and later obtaining a deeper understanding of the origins of these properties at lower levels, as in the example of chemical bonds based on quantum mechanics. The limitation on precise simulations also does not prevent construction of useful *models* of complex systems; we just have to be very careful to distinguish the idealized models from the actual systems. In the context of putative computer intelligence, I see no reason to expect that any idealized brain model can achieve consciousness. Rather, consciousness may require accurate representation at many, or perhaps all, scales, an achievement that may be fundamentally impossible for the reasons outlined above.

10.4 ULTRA-INFORMATION

We have seen that "information" is a term that has very different meanings depending on context. *Entropy* refers to ordinary information that is normally inaccessible to measurement but may become accessible with appropriate observations. The dynamic state of our fanciful baseball in chapter 9 was determined by perhaps 500 bits of accessible information, while its entropy was estimated as something like 10^{25} bits. This distinction is indicated by figure 10-2, where ordinary information is pictured as consisting of accessible information and inaccessible information or entropy. In order to bridge the gap between the languages of the physical and mental worlds but, at the same time, avoid some of the hazardous linguistic traps, we consider an additional distinction between informational categories as indicated in figure 10-2. In this picture *ordinary information* is assumed to be part of a broader category called *ultra-information*. The specific definition I adopt here is quite broad—*ultra-information is that which distinguishes one thing from another.*

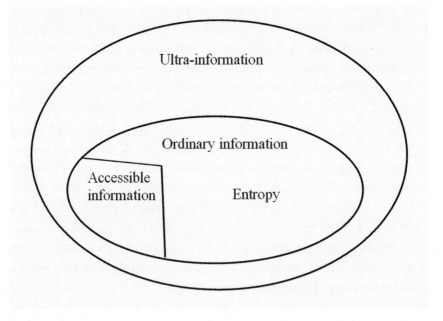

Fig. 10-2.

The sensory input from the physical world that enters the minds of human observers involves several kinds of ordinary information transfer and processing, including external physical phenomena like light and sound, as well as internal physiological events like action potentials, cellular integration of synaptic inputs, chemical-transmitter actions, and so forth. Here we address the hard problem of consciousness by posing questions about the possible existence and nature of something well beyond ordinary information, that is, ultra-information, defined broadly to include ordinary information, unknown physical processes, and consciousness, as pictured in figure 10-2. Thus, thoughts, emotions, self-awareness, memory, and the contents of the unconscious are, by definition, categories of ultra-information whether or not these mental processes also consist of ordinary information. Given this general conceptual framework, several basic questions come to mind. First, is any of this so-called ultra-information associated with some mysterious field or "force" that operates outside the realm of science? While such a possibility cannot be ruled out, such extra-science speculations can lead us nowhere other than this book's immediate end. Thus, we will avoid this abrupt ending by sticking with interpretations that are consistent with mainstream science. Such a limitation does not, however, preclude fields that are currently unknown but do not seem to conflict with known physical laws.

Another immediate question that comes to mind is whether any separation actually occurs between the two ellipses in figure 10-2, which distinguish ordinary information from ultra-information. Does such a category of information actually exist that is fundamentally far more inaccessible to observation than ordinary entropy, that is, *forever and absolutely unknowable*? To this question we offer an unequivocal yes. We will soon see that certain types of hidden (unknowable) information are part and parcel of mainstream physics. Such hidden information takes on a variety of different labels and interpretations, but its existence is just as well-established as any other scientific principle. Any attempt to include consciousness in the same informational category is likely to be controversial; however, the broad definition of ultra-information given above leaves us with plenty of wiggle room. All we are saying is that such a loose connection between consciousness and modern physics is theoretically possible; it need not violate known physical laws.

This discussion about the *existence of the unknowable* may seem rather esoteric; however, we are concerned every day with an unknowable entity that is widely believed to exist; we call this entity *the future*. As discussed in chapter 9, prediction of the future behavior of complex systems is always limited by uncertainty in initial conditions, boundary conditions, and interaction rules. This limitation is fundamental to both classical and quantum systems; it applies to financial markets, humans, climate, and other physical systems. Our predictions of the future must always be of a probabilistic nature even when we hide our ignorance behind the veil of deterministic language.

According to the theory of relativity, even much of our *past* is fundamentally unknowable. The star Betelgeuse is (or was) located 642 light-years from Earth. Ordinary information about any event that occurred near this star anytime in the past 642 years is, for humans, hidden by a thick fundamental veil that cannot be lifted by human technology. For all we know, Betelgeuse no longer exists; perhaps it exploded in a supernova or was swallowed by a black hole during the year of Christopher Columbus's birth (1451). Because our universe imposes a strictly enforced speed limit (the speed of light) on ordinary information transfer, we cannot know anything of Betelgeuse's condition at the time of Columbus's birth until the year 2093. According to established physics, such events exist but are fundamentally unknowable.[9] While this kind of unknowable information, based on the universal speed limit, can apparently have no effect on us in the present, the field of quantum mechanics involves hidden information that can influence us, apparently without delay.

10.5 QUANTUM MECHANICS: SCIENCE AT SMALL SCALES

Quantum mechanics is the science of small-scale systems—molecules, atoms, and below; its accuracy has been verified in thousands of experiments. Nevertheless, quantum concepts remain largely unappreciated by most nonphysicists (including many scientists) and run counter to everyday experiences, seeming more like magic than like science. For many, quantum mechanics seems too abstract and remote from daily life to be of interest; most persons have hardly given it a second thought. But quantum-mechanical footprints are everywhere, quite evident in the natural world as well as in human technology, touching our lives on a regular basis. The well-established and widespread experimental verifications of weird quantum effects provides some intriguing *hints* of possible quantum connections to mind, suggesting that the idea be taken seriously if, as purported here, the hard problem of consciousness lies within scientific purview.

While several prominent scientists have proposed deep connections between quantum mechanics and consciousness, many scientists view such ideas with open contempt—skeptics view the speculative links simply as vague notions that since both topics are mysterious, perhaps they are somehow related. But even if the skeptics are correct, the grand conceptual leap from classical to quantum thinking may provide some feeling for how far brain science may eventually stray from current ideas. The resulting humility may make us especially skeptical of attempts to *explain away* (with tautology) observations that fail to merge easily with common notions of consciousness—multiple consciousnesses associated with single brains, group consciousness, the role of the unconscious, hypnosis, and so forth.

If nothing else, quantum mechanics provides important epistemological lessons for brain science and philosophy. In simpler terms, the weird worldview provided by modern physics (relativity and quantum mechanics) should give pause to those philosophers and scientists who would trivialize the hard problem of consciousness simply because they find it "unscientific" or even occult-like. Too often their idea of "unscientific" is based on the outdated notions of nineteenth-century physics. But irrespective of this controversy, there is no question of the essential contributions of quantum mechanics to consciousness in one important context—quantum mechanics underpins biochemistry, which clearly influences or even controls consciousness in a myriad of ways. Thus, by following the implications of the multiscale conjecture down to molecular and sub-molecular scales, we are led unavoidably to consider quantum-mechanical issues.

The field of quantum mechanics is far too large and dependent on mathematics to be appropriate for treatment in this book. We can, however, touch on several quantum ideas related to ultra-information. But we will proceed with caution—quantum science is so weird that it is has opened the door to many speculations that run the gamut from plausible ideas to baloney. To keep the weird ideas in check, we employ table 10-1, which outlines established scientific concepts in the first three columns. The fourth column lists "serious" speculative ideas, defined here as positions apparently consistent with mainstream science and considered plausible by more than a few bona fide physicists.[10] We have no need to identify or consume the actual baloney, which is omitted from table 10.1.

Table 10-1. Established (I–III) versus Speculative (IV) Physics.

I Classical Physics	II Quantum Mechanics	III Relativity	IV Speculative
• Newton's laws • Chaos • Complex systems • Circular causality • Electromagnetic waves in empty space • Energy conservation • Entropy increases in closed systems	• Wavefunction yields statistical information • Uncertainty principle limits quantity of information • Quantum tunneling in modern technology • Light is wave or particle depending on experiment • Entanglement • Nonlocal influences over large distances (spooky)	• Einstein 4D space-time • Mass is stored energy • Space limits speed of information • Time dilation • Black holes • Dark energy • Dark matter	• Wavefunction collapse • Bohm mechanics • Many worlds • Infinite multiverse • String theory • Worm holes • Information is fundamental • Quantum influences on consciousness

Quantum mechanics reveals a holistic character of the universe that is entirely foreign to classical thinking. The scientific observer and the observed system are joined in intimate ways governed by an "information field" called the *quantum wavefunction*. Particles also seem to reside in a kind of quantum limbo or alternate realm of reality with undefined properties until observed with scientific instruments. Particles may jump between locations, evidently without ever occupying the space in between, *as if* reality itself consists of successive frames of a motion picture. This background of physical principles and information perhaps hints of links to the hard problem if consciousness is viewed as a special, expanded category of ultra-information processing.

The actual practice of quantum mechanics is carried forward in two distinct stages.[11] One stage involves the quantum wavefunction; the second stage concerns the specific kind of experimental study, including both experiment preparation and choice of measurement options. In the traditional (Copenhagen) interpretation, apparently now rejected by many physicists, such experiments involve a strange process called *collapse of the wavefunction*. In very simple terms, I outline this process as follows: Sometimes a particle can be located in two or more places at the same time, but the act of measurement causes it to "collapse" to one location or the other. One cannot be sure just where it will appear. Fortunately, this behavior only occurs at microscopic scales. When you park your car, you can be quite sure that it will occupy only one parking place at a time. You need not worry about "car uncertainty" or hundreds of tickets for illegal parking.

To adopt still-another metaphor, the "easy problem" of quantum mechanics involves wavefunction calculations, whereas the "hard problem," commonly known as the *measurement problem*, involves onto-logical interpretations of quantum observations. The easy problem of quantum mechanics involves solutions of a deceptively simple-looking partial differential equation (Schrödinger's equation), which provides unambiguous statistical information about a collection of particles. We have no such convenient mathematical framework at our disposal to address the hard (measurement) problem of quantum mechanics.

10.6 ULTRA-INFORMATION AND THE RQTC CONJECTURE

The physical laws that govern our universe are laws about information. The second law of thermodynamics requires that inaccessible information (entropy) in closed systems must increase over time. The theory of relativity limits the speed at which information may be transmitted through space. Quantum mechanics provides statistical information about focused experimental observations but imposes strict limits on what can ever be known. If we view consciousness as some higher kind of information processing, the possible existence of some sort of fundamental link between the physical and mental worlds does not appear so unreasonable, even though we have little or no idea of how this relationship might work. Here we tag such a putative connection the *RQTC conjecture*, standing for relativity, quantum mechanics, thermodynamics, and consciousness. This conjecture does not presuppose anything about the nature of such connection.

Now I will briefly discuss several entries in table 10-1 that appear consistent with our ideas about ultra-information. The uncertainty principle (column II) places fundamental limits on the precision with which certain pairs of physical properties, known as complementary variables, such as the location and velocity of a particle, can be known simultaneously. For example, as discussed in chapter 9, the total information content of a baseball, consisting of macroscopic (accessible) information plus entropy (inaccessible information) is finite. Any other information, if it exists at all, is absolutely and forever hidden; in the language of this chapter, this is ultra-information.

Consider one critical consequence of the uncertainty principle. Every atom consists of a positively charged nucleus surrounded by negatively charged electrons. Classically, charged particles lose energy when accelerated, as occurs in orbital motion. This raises a fundamental question—why don't the attractive forces between charges cause electrons to fall into the nucleus after they have lost speed? The apparent answer is that such action would fix the location of each electron within a very small space, thereby requiring the uncertainty in speed to be very large, thereby implying that electrons would fly out of the atom altogether. But this superficial picture lacks self-consistency; a stable atom actually requires a compromise in which electrons jiggle here and there with moderate uncertainty in both speed and location within the atom. This picture emphasizes the central importance of the uncertainty principle even when no experiment occurs. Electrons are evidently *forbidden by nature* to acquire precise locations and speeds at the same time, otherwise, groups of electrons, protons, and neutrons could never stick together to form atoms. *If not for the uncer-*

tainty principle providing limits on information, atoms, molecules, base-balls, and humans would not exist.

Another weird aspect of quantum mechanics is wave-particle duality. Light, electrons, and other small particles can behave like particles or waves depending on how they are measured, as indicated in column II of table 10-1. Even fairly large-sized molecules like the buckminsterfullerene or "bucky-ball," which consists of sixty carbon atoms, have been shown to exhibit such wave-particle duality. For example, turning on a sensor in one part of an experimental device can cause an *electron wave* in some other part of the device to switch its character to an *electron particle* through some unknown informational influence.[12]

To emphasize just how strange this is, consider a fanciful metaphor—our hero, Bill, begins to enter the front door of his house, which has a back-door camera that is initially turned off. Inside the house, Bill's wife, Alice, quickly turns on the camera just before Bill can get inside. The camera indicates that nobody is entering the back door. As a result of this seemingly innocent action, the man Alice sees coming inside the front door is her secret lover, Bob, rather than her husband, Bill! In other words, when Alice (or some recording device) obtains information about visitors entering the back door, this act somehow influences the identity of persons entering the front door.

From a classical perspective, this kind of weird identity transfer makes no sense at all. How can a particle be localized to a small space when some remote sensor is on, but be distributed over a large region like a wave when the distant sensor is turned off? This question has been the subject of an ongoing debate among physicists since the 1920s, but some sort of infor-mational-like influence must be involved. All of the proposed explanations are weird; none make any sense based on our everyday experiences. But, regardless of which strange interpretation is valid, we seem to be on solid ground by including such well-established influences of the measuring device on the measured particle in the category of ultra-information. In the context of our crude metaphor, the seemingly innocent act of switching on the rear-door camera has somehow magically changed Bill into Bob.

In the traditional (Copenhagen) interpretation of quantum mechanics, some of an unmeasured particle's properties are, in a certain sense, not real; properties like position and speed are realized (created) only in the act of measurement. An alternate view, called *Bohm mechanics*,[13] says that while the unmeasured particle is quite real, its measured properties depend on multiple influences that act over arbitrarily large distances from the measurement site, *nonlocal influences* that may act at greater-than-light speed or even infinite speed. In this view, a particle's location is caused by a vast collection of distant events, potentially in other parts of the universe. While Bohm mechanics itself is controversial, the nonlocal influences are not; they constitute an essential part of mainstream quantum mechanics, as indicated in column II of table 10-1. I'm sorry if I have lost you with the discussion above. But, don't feel too bad; many prominent physicists have repeated the dictum, "nobody really understands quantum mechanics." We conclude that the quantum physicist is a little bit like some kid who has obtained a magic wand from a master magician; the wand works every time, but the kid has no idea of how it works.

Einstein famously hated the nonlocal aspect of quantum mechanics, employing the derisive description, "spooky action at a distance," but in this case Einstein has been proved (partly) wrong. While relativity theory prohibits the transmission of information above the speed of light, no such restriction seems to apply to influences that carry no *ordinary information*; that is, information that can, in principle, be decoded by observers. Here we place such established spooky influences in the category of *ultra-information*.

Although quantum mechanics has been fully tested with a myriad of disparate experiments, the results are subject to conflicting ontological interpretations. Quantum-mechanical questions may be divided into instrumental questions (expressed in terms of either actual or thought experiments) and purely ontological questions (for which no experimental tests are currently known). Ontological questions, which concern the nature of reality, continue to be argued in both the scientific and general literature, and new experiments may be proposed to move ontological questions into the instrumental category for resolution. These ideas are clearly discussed in nonmathematical language in many books[14] and will not be repeated here, but the bottom line is that all interpretations of quantum mechanics are weird or occult-like when considered from our everyday, classical viewpoints. Readers holding fast to classical worldviews should not expect to find intuitively pleasing explanations of quantum mechanics; scientists smarter than us have repeatedly failed in this quest over the past century.

10.7 FIELDS AND DYNAMIC INFORMATION PATTERNS

The implications of the *multiscale conjecture* at small scales have caused us to speculate on possible relationships between consciousness and *ultra-information*. This esoteric (but well-established) entity leads us naturally back to the subject of *fields*. Thus far, we have been satisfied to simply equate the label "field" to *dynamic-information pattern*. But, in order to facilitate a broader picture, we now consider the issue more closely—just what are "fields"? Earlier, we described chaos in boiling pots of soup, so we may as well add a metaphorical oven to our fanciful kitchen.

When the oven is turned on, the kitchen air is heated. If the air temperature at a number of locations near the oven is measured at multiple times, this spatial-temporal pattern can be expressed by some sort of list or by a mathematical function $T(x, y, z, t)$. This symbolism simply means that the temperature T depends on the location of the thermometer (its coordinates x, y, z) and the time of measurement, t. Temperature, pressure, mass density, air speed, and so forth are all *fields* based on some underlying structure. Temperature is due to the random motion of air molecules; air speed is their directed motion. We can associate both information and energy with these fields—the temperature-field energy in a given volume of air is simply the total kinetic energy of the air molecules contained therein; the associated information is mostly entropy.

The representation of a temperature pattern as a *field* is mainly a matter of mathematical convenience, a very useful, but for our current purposes, rather-mundane exercise. Fields often represent large-scale properties of some underlying small-scale activity, as in the case of molecular interactions producing temperature. However, we are here interested in something much more interesting, namely the more-abstract fields that have no known underlying *carrier*, a structure to "carry" the field. This important distinction has its roots in nineteenth-century physics, which focused for several decades on finding an underlying structure for *electromagnetic fields*. Many attempts were made to find the putative supporting carrier called the *ether*. After all, sound waves, water waves, seismic waves, and so forth involve physical structures in motion, but experiments failed to reveal evidence for the ether. This outcome, together with Einstein's special theory of relativity in 1905, eventually *forced* abandonment of the idea that electromagnetic fields require a physical medium in which to propagate.

I emphasize the word "forced" because many scientists of the time did not surrender their mechanical viewpoint easily—some died refusing to accept the implied "occult-like" picture of physical reality. That particular resistance is now long past; scientists now regard electromagnetic fields as something *much more profound* than the early physicists did . Electromagnetic fields retain an abstract quality—*the fields are defined only in terms of the observable effects that they produce on charges and currents*. Even though they lack an underlying physical structure, electromagnetic fields carry both information and energy, features contained *in the field itself.* This critical property allows transmission of the sun's energy over millions of miles of empty space, making life on Earth possible.

Let's reflect a bit on this strange behavior—just what is an electromagnetic field? Sure, we can draw wiggly lines and write down the fancy mathematics of fields, which make accurate predictions of observable effects, but we cannot say too much more about its wave-like nature. A radio or cell-phone transmitting station at some location X generates coded current patterns in a broadcasting antenna, thereby generating an electromagnetic field that becomes "detached" from location X and propagates away in all unblocked directions. At a second location, Y, perhaps your cell phone, this electromagnetic field induces current in the local antenna in the same frequency band(s). In short, you jiggle your electrons at X and my electrons at Y jiggle a little later in a predictable manner, even in empty space. If my radio or cell phone is tuned in, information is received. With this story in mind, note that table 10-2 lists a number of phenomena that may be treated as "fields," together with the sensors or "antennae" that respond to the matching field. The special fields listed in italics are distinguished from the others by their lack of any known underlying structure. We only know of their existence indirectly, as a result of the observed actions of the matching antenna. The bottom entry refers to a highly speculative idea outlined in the following section.

Table 10-2. Fields and Matching Antennae.

Fields acting top–down	"Antenna" that responds to field
Social systems	
large-scale culture	social networks
economic depression	poor people
Physical systems	
temperature	thermometers
electromagnetic field	*charge in cell phones*
quantum wavefunction	*elementary particles*
dark energy	*space-time?*
dark matter	*ordinary mass?*
Physiological systems	
sound	inner-ear membrane
odor	olfactory system
light	*visual system*
magnetic field	*magneto-reception in birds*
electric field	*electro-reception in electric fish*
Brain systems	
global synaptic action field	neural networks
large-scale network	small-scale embedded networks
ultra-information	*brains?*

Early scientists who found the idea of electromagnetic waves propagating through empty space to be "occult-like" must have had their parochial worldviews completely blown away in the early twentieth century. Practicing engineers had begun to regard the electromagnetic field as a real physical entity, based largely on the incredible accuracy of Maxwell's equations in predicting the behavior of charges and currents in macroscopic systems. But this worldview was soon to encounter a serious challenge in experiments conducted at microscopic scales. Einstein reaffirmed the validity of Maxwell's *wave* equations in his 1905 paper on special relativity. Amazingly, in the very same year, he also published a paper offering a very *anti-wave* explanation for the photoelectric effect—if you shine a light on a metal surface, electrons are emitted. Based on the wave nature of light, one would expect the kinetic energy of emitted electrons to depend on the *intensity* of light, which is proportional to the squared magnitude of the electromagnetic field, but this prediction is dead wrong. Einstein showed that the experimental observations of emitted electrons could be explained only if light was assumed to be composed of little packets of energy, now called *photons*. The photon and emitted electron energies are proportional to *wave frequency*; that is, the *color* of the light. Blue light will do the trick; red light will not. Einstein's Nobel Prize in 1921 was awarded to him for his work on the photoelectric effect, not his works on relativity, which were more controversial at the time.

The photoelectric effect led scientists to adopt the "occult-like" concept of *wave-particle duality*, initially only with great reluctance— light and (later) particles with mass simultaneously possess the properties of both waves and particles, each property manifested according to the experimental circumstances. Scientists were *forced* by the experimental evidence to accept this weird view, but now the new generations of physicists have become so accustomed to this picture that the tag "occult" is rarely, if ever, employed. If you are a budding scientist, you better just get used to it—light and other electromagnetic fields, as well as small particles with mass, can behave *as if* they are waves, or they can act *as if* they are particles. The wave and particle descriptions are complementary models of physical reality. How can this possibly be? If you find this idea difficult to grasp, take comfort in what Einstein said in 1954—"All these fifty years of conscious brooding have brought me no nearer to the answer to the question, 'What are light quanta?' Nowadays every Tom, Dick and Harry thinks he knows it, but he is mistaken."[15]

The quantum wavefunction, which is determined partly by the experimental apparatus and choice of measurement, acts on elementary particles. Other kinds of fields act on macroscopic objects. *Dark matter* is hypothetical mass whose existence is inferred from the observed motion of stars and galaxies. In addition, some sort of *dark energy* appears to permeate space, as it provides an explanation for the accelerated expansion of the universe. Surprisingly, dark energy plus dark matter appear to make up 96 percent of the mass-energy of the observable universe, with visible matter like stars and intergalactic gas accounting only for the remaining 4 percent.[16] If dark energy and dark matter have genuine physical properties, associated *dark information* must also exist. I am not claiming that such dark information has anything to do with consciousness. Rather, the discovery of these strange new fields serves to remind us that still more fields, as yet unknown, may be found; some might even have links to consciousness. I offer no support for this conjecture, only the glib (and not altogether convincing) retort, *absence of evidence is not evidence of absence.*

10.8 A "CRAZY" RQTC IMPLICATION: TOP-DOWN BRAIN INFLUENCES

We have cautiously followed the implications of the multiscale conjecture down to very small scales, challenging the notion that dynamic patterns of physical activity associated with consciousness must terminate at the cellular level. This conceptual approach questions whether modern physics can shed light on the fundamental challenge of consciousness in a manner that is impossible using only classical concepts. We label such connection to modern physics the *RQTC conjecture*, independent of its actual nature. I have promised to limit speculations to those ideas that do not obviously violate known physical or biological laws, a criterion meant to distinguish responsible speculations from outright baloney. While I fully intend to honor this standard, we should recognize that some ideas that today's philosophers and nonphysical scientists seem to view as "unscientific" are actually embraced by mainstream modern physics. A more-plausible modern scientific position is that studies of consciousness should be prepared to employ a full complement of scientific tools, including relativity, quantum mechanics, and thermodynamics. This viewpoint does not prejudge the relevance of such modern tools; it only says that we should allow them a test run.

Despite all the caveats and disclaimers in the paragraph above, this section may test the patience of some skeptical readers by speculating on putative strange implications of the RQTC conjecture, including unknown top-down influences acting on brains, perhaps in the category of some sort of universal ultra-information, as suggested in the bottom row of table 10.2. In stark terms, we now speculate on the idea that *the brain may act as kind of antenna that interacts selectively with some external entity*. I am not advocating this idea, nor do I offer supporting evidence. As a scientist having spent a career studying relationships between the brain's dynamic patterns and mental states, my intuition leans toward anti-dualist positions. Yet I also favor a frank admission of our profound ignorance of the origin of conscious awareness.

A common suspicion of skeptics is that dualism or RQTC proponents are searching for ideas to support preexisting or ingrained religious views. Let me again assure readers that I have absolutely no motivation to support any particular religious belief, as discussed in note 4 of chapter 2. Rather, I intend to forestall implicit censorship by religious fundamentalists, atheists, or any other group. Some skeptics, for example, may fear opening a philosophical back door that allows religion to sneak in and misdirect logical arguments. Skeptics often raise valid points about the need to avoid religious bias in science, but my reasoning is this—if consciousness studies lie within the purview of legitimate scientific enquiry, RQTC must be seriously considered if only because classical physics cannot even begin to explain consciousness. In broader terms, I have come to believe that the study of consciousness is essentially a study of the nature of reality, and no serious thinker should doubt that modern physics deals with important aspects of physical reality.

How might brains act like antennae? In table 10.2 we see that physics has identified several fields that have no known underlying structure—electromagnetic fields, dark energy, and the quantum wavefunction, for example. The first two fields carry both information and energy, the wavefunction carries only information. We know of the existence of these fields only through their influence on matching "antennae." Electrons, for example, have certain fundamental properties like *charge* and *spin*, which serve as sensors or antennae of electromagnetic fields. I am not suggesting that these fields have anything to do with consciousness. However, for all we know, our universe may contain other fields that are as yet undiscovered; these may fall into our ultra-information category. Over the past century or so, most scientists have regarded mass and energy as nature's primary actors on the universal stage, with information allowed only a secondary role. However, some modern scientists advocate a radically different view, one that elevates information to the fundamental entity underlying all of physical reality, implying the conceptual hierarchy: *information → laws of physics → matter*.[17]

Why should one take this idea of *information-as-fundamental* seriously? A short answer, addressed in chapters 9 and 10, is that the known physical laws, relativity, quantum mechanics, and thermodynamics, are all laws about information. The wavefunction of a system of quantum particles encapsulates all that is known about the system; it essentially acts as an information field. The imagined fundamental information or ultra-information might be embedded in our existing space-time (as the usual quantum wavefunction), or ultra-information might create space-time itself. What are the implications of this revolutionary new paradigm? For one thing, it is bound to be controversial. For some scientists this may be likened to opening Pandora's box or at least a large can of worms, releasing all kinds of wild ideas about the origins of consciousness, implications for religious beliefs, mysticism, and so forth. Others, especially those interested in the hard problem of consciousness, may be more receptive to the new ideas. Minds and information processes appear, in the information-based picture, to be integral parts of our universe rather than accidental products of evolution (as assumed by many). If information is fundamental in our universe, then maybe consciousness is fundamental.

All this talk of fundamental information or ultra-information is, of course, quite speculative. One cannot discount the enormous success of relativity, thermodynamics, and quantum mechanics in describing our physical world. The proposed new informational paradigm currently provides no equivalent mathematics to predict the outcomes of specific experiments or develop new technology. Large parts of this paradigm are currently nonfalsifiable, leaving them beyond scientific purview, at least for now. Does the brain create the mind, or is something else like ultra-information involved? Your guess is as good as mine.

10.9 SUMMARY

Our quest to better understand consciousness consists of betting on a long shot because this deep mystery is profound and possibly unsolvable. But, despite the severe limitations of our acknowledged ignorance, we can perhaps shed a little light on the mystery. In so doing, we also explore some speculative ideas that they may seem preposterous to some; however, our speculations do not appear to violate established physical or biological laws.

Our main messages revolve around the *multiscale conjecture*, which proposes that consciousness is "coded" at multiple spatial and temporal scales. This implies that multiscale dynamic patterns are necessary for consciousness to occur—top-down and bottom-up interactions across scales (circular causality) are essential to consciousness. This conceptual framework is consistent with both materialist and nonmaterialistic philosophical positions on consciousness. At minimum, consciousness seems to require systems that are at least as complex as ordinary life, and all known life is based on nested multiscale structure and interactions. The multiscale conjecture opposes philosophical positions that appear to trivialize consciousness by viewing brains as simple systems.

The multiscale conjecture is employed here with two possible directions in mind—first, as a stand-alone partial "explanation" of consciousness signatures, independent of materialism/dualism issues and the hard problem. But, in addition, the multiscale conjecture may provide a "missing link" connecting brain information patterns to a lesser or perhaps even nonmaterialist conceptual framework underlying the hard problem. In either context, we are faced with the fundamental question, If the multi-scale conjecture is correct, where do we make the small-scale cutoff below which physical structures may be treated simply?

In order to approach the issue of multilevel consciousness, we note that memory consists of several phenomena closely associated with consciousness. Short-term, long-term, immunological, and genetic memories require information storage on progressively longer timescales. Short-term memory is associated with dynamic patterns and cell assemblies. Immuno-logic and genetic memories are stored at smaller spatial scales, including cellular and molecular scales. The usual sharp distinction between short- and long-term memories masks a continuum of different processes involving interactions across both time and spatial scales.

Ordinary life consists of structures at many scales arranged in nested hierarchies. Here we focus on the cellular and lower levels—molecules, atoms, electrons, and ultimately the Planck scale, where space-time itself is believed to end. Larger-scale systems *emerge* from interactions at smaller scales and display novel features that are entirely missing from the smaller-scale entities. The macromolecules of living cells are believed to obey the principles of quantum mechanics. But, this connection across spatial scales may have led some scientists and philosophers to embrace a major fallacy—namely that biochemistry is simply applied physics. While no features of macromolecules are known to violate quantum mechanics, this result does not imply that scientists can *predict* large-scale properties from interactions at the atomic level. Rather, biochemistry relies on the emergent properties of macromolecules, resulting in no small part from the environmental conditions in which macromolecules form. The environmental conditions may even encompass unknown fundamental organizational principles, perhaps some informational features. But, in any case, biochemistry is clearly far more than just applied physics.

"Information" has very different meanings depending on context—ordinary information consists of both accessible information and inaccessible information, or entropy. Here we propose a broader category that includes ordinary information—*ultra-information*, which is defined as that which distinguishes one thing from another. Sensory input from the physical world involves several kinds of ordinary information—light and sound plus internal physiological events like action potentials, chemical transmitter actions, and so forth. Ultra-information includes ordinary information, unknown physical processes, and consciousness. Thus, thoughts, emotions, self-awareness, memory, and the contents of the unconscious are, by definition, categories of ultra-information whether or not these mental processes also consist of ordinary information.

I offer no proof of any direct connections between quantum mechanics and consciousness; however, weird quantum effects provide some intriguing *hints* of possible quantum links. Quantum mechanics proceeds in two stages—the first involves the mathematics of the quantum wavefunction. The second stage concerns the actual experimental system, including both experiment preparation and choice of measurement options. The "easy problem" of quantum mechanics involves wavefunction calculations, whereas the "hard problem," commonly known as the *measurement problem*, involves ontological interpretations of possible quantum observations.

The physical laws that govern our universe are informational laws. Thermodynamics requires that inaccessible information (entropy) in closed systems must increase over time. Relativity limits the speed at which information may be transmitted through space. Quantum mechanics provides statistical information about experimental observations, but it imposes strict limits on what can ever be known. If we view consciousness as some higher kind of information processing, the possible existence of some sort of fundamental link between the physical and mental worlds appears plausible, even though we have little or no idea of how this link might work. This putative connection is the *RQTC conjecture*, standing for relativity, quantum mechanics, thermodynamics, and consciousness, independent of the actual means by which this connection occurs.

Speculation about the nature of *ultra-information* leads us to consider the subject of fields in depth; fields are essentially dynamic-information patterns. Some fields, like temperature distributions, are based on an underlying molecular or other structure. Others, like the quantum wavefunction, dark energy, or the electromagnetic field, have no known underlying structure; they simply exist within space-time. Electromagnetic fields retain an abstract quality—the fields are defined only in terms of the observable effects that they produce on charges and currents. Electromagnetic fields carry both information and energy; quantum wavefunctions carry information but not energy. Each of these features is *contained in the field itself.*

The multiscale conjecture is followed down to very small scales, challenging the notion that dynamic patterns of consciousness must terminate at the cellular level. Can modern physics shed light on the fundamental challenge of consciousness in a manner that is impossible using only classical concepts? In this book, speculations are limited to those that apparently enjoy consistency with known physical and biological laws, hopefully distinguishing responsible speculations from outright baloney.

Skeptics may suppose that proponents of dualism or the RQTC conjecture are attempting to support ingrained religious views. Some may fear opening a philosophical back door that allows religion to sneak in and misdirect logical arguments. The skeptics raise valid points; however, my justification for speculation is this—if consciousness studies lie within the purview of legitimate scientific enquiry, RQTC must be seriously considered if only because classical physics cannot even begin to explain consciousness. We end this book with the highly speculative idea involving some unknown ultra-information that acts top-down on brains such that brains act as "antennae" interacting with this external entity. No evidence for this "crazy" idea is supplied; we can only say that this or some similar dualistic picture is theoretically possible.

GLOSSARY

abiogenesis: The natural process of life emerging from nonliving matter like organic chemicals.

action potentials: Electrical signals that travel along axons.

AI (artificial intelligence): Intelligence exhibited by machines or software.

alpha rhythms: EEG signals with relatively large frequency components in the 8 to 13 Hz range.

ALS (amyotrophic lateral sclerosis): Lou Gehrig's disease; may lead to the dreaded *locked-in state*.

Alzheimer's disease: A chronic brain disease typically involving a very gradual loss of normal consciousness.

amplitude: The magnitude of an oscillatory waveform or signal.

amyotrophic lateral sclerosis: See *ALS*.

analogue: A system that has many properties in common with some different system, implying a stronger relationship than a metaphor or a cartoon.

antigen: A foreign body that evokes an immune response.

aphasia: A group of language disorders often caused by brain damage.

artificial intelligence: See *AI*.

astrocyte. A (non-neuron) cortical cell believed to participate in cortical processing.

autism: A disorder of brain development characterized by impaired social interactions and communication, as well as restricted and repetitive behaviors. Parents usually notice signs in their children by age two or three.

axon: The output fibers from neurons that end in synapses.

backward masking studies: Suppose two stimuli are presented to a subject. The second stimulus is presented less than 500 milliseconds after the first. The second stimulus may "mask" conscious awareness of the first stimulus.

basilar membrane: An inner-ear structure that responds selectively to specific sound frequencies.

binding by resonance: Cross-frequency coupling caused by resonant interactions between distinct (semi-autonomous) subsystems, including brain networks.

binding by synchrony: A process by which different features of an animal's world are represented by distinct brain networks that become functionally connected by synchronous rhythms. A subcategory of binding by resonance.

binding problem: The issue of how widely distributed neural activities give rise to a single integrated behavior and consciousness.

black hole: A region of space-time with such a strong gravitational field that nothing, not even light, can ever escape from the inside, the ultimate fate of stars above a certain mass.

blindsight: Subliminal processing of visual images that allow unconscious information storage.

Bohm mechanics: A nontraditional ("non-Copenhagen") interpretation of quantum mechanics that explicitly exposes its inherent nonlocal character.

bottom-up interactions: The influences of small-scale systems on large and emergent systems.

boundary condition: In general use, the external environmental conditions within which some localized system evolves. In mathematical physics, the physical constraints placed on variables on the surface (boundary) surrounding the local system under study.

brain dynamic behavior: The patterns of various kinds of activity that change over both time and brain location, analogous to weather patterns.

brain-friendly: An idealized brain idea or model that appears compatible with genuine neuroscience.

brain rhythm: The oscillation of brain electric potential patterns, normally recorded with electroencephalography (EEG), electrocorticography (ECoG), or local field potential methods (LFP).

Broca's area: Major speech production region located near the intersection of the temporal and frontal lobes.

Brodmann region: One of fifty-two areas of the cerebral cortex defined by cell organizations.

Capgras syndrome: Delusion that friends or family members have been replaced by identical-looking imposters.

callosal axons: Nerve fibers that connect the two brain hemispheres through the corpus callosum.

CCC (*cortico-cortical column*): A cortical-tissue column defined by the spatial spread of subcortical input axons that enter the cortex from the white matter below; it contains about ten thousand neurons within its diameter of approximately 0.3 mm.

cell assembly: A group of neurons acting as a single system over some time interval, essentially a "neural network" defined broadly.

cerebral cortex: A folded structure forming the brain's outer layer, believed to be responsible for most mental activity.

cerebrospinal fluid: See *CSF*.

C-factor: Some essential aspect of humanness missing in philosophical zombies (or "p-zombies").

chaos: The behavior of dynamic systems that are hypersensitive to initial and boundary conditions. In essence, microscopic information "infects" macroscopic properties. While the present determines the future, the approximate present does not approximately determine the future.

Christmas-tree brain: A metaphorical tree with lights corresponding to active brain regions.

circular causality: The combined bottom-up and top-down interactions across spatial scales.

classical physics: Everything known about the physical world before 1905.

coherence: The consistency of the phase-difference between any two signals, with each signal normally composed of multiple frequencies. Expressed as a (squared) correlation coefficient for each frequency component with magnitude always between zero and one. Coherent signals may or may not be synchronous (zero phase difference).

coma: An unconscious state in which one cannot be awakened or respond to stimuli.

complementarity: Aspects of reality that have alternate descriptions in different contexts; used in both complex classical systems and quantum systems.

complexity science: The study of complex systems composed of many interacting parts. Such systems exhibit global (large-scale) behavior that cannot be easily explained in terms of their smaller parts.

connectome: The brain's wiring diagram over some range of spatial scales.

consciousness: The state of awareness of an external environment or of something within oneself. The ability to experience or feel at some level. May or may not include self-awareness.

consciousness signals: Generally, measured correlates of various mental states. Sometimes restricted to conscious versus unconscious states.

continuum mechanics: A branch of science and engineering that deals with the dynamic behavior of continuous media rather than discrete entities like network nodes.

Copenhagen interpretation of quantum mechanics: Long considered the mainstream view, but now apparently rejected by most physicists. An early attempt by Bohr and Heisenberg to reconcile the apparent dualism of "wave" and "particle" in a way that humans can understand.

correlation: A predictive relationship between two variables.

cortico-cortical column: See *CCC*.

cortico-cortical fibers: The white matter axons that interconnect different parts of the cerebral cortex to itself.

cross-frequency coupling: Functional connections between rhythms of different frequencies.

C-scale. An imagined special scale at which consciousness is encoded; the single neuron and molecular scales are two candidates. In this view, consciousness signatures observed at other scales are mere by-products of C-scale dynamic behavior. This idea is discounted by the multiscale conjecture.

CSF (*cerebrospinal fluid*): Fluid in the brain and spinal cord that carries important chemical messages to neurons via neuromodulators.

cybernetics: The scientific study of control and communication in animal or machine.

dark energy: An unknown kind of energy field believed to permeate all of space and that tends to accelerate the expansion of the universe.

dark matter: An unknown substance believed to account for a large fraction of the matter in the observable universe.

delta rhythm: EEG oscillations with frequencies in the approximate range of 1–4 Hz.

dementia (*senility*): A broad category of brain diseases that cause long-term reductions in thinking and memory abilities.

determinism: A range of positions in which future events are fully or mostly determined by current conditions.

diffusion tensor imaging: See *DTI*.

DTI (*diffusion tensor imaging*): An application of MRI technology used to measure preferred directions of water diffusion in brain voxels, suggesting matching directions of major fiber (axon bundle) tracts.

dualism: The view that the mental and physical realms are distinct and separate aspects of reality.

easy problem: Finding relationships between states of mind and various electrical, chemical, metabolic, or other measures of brain activity called *signatures of consciousness*.

EEG (*electroencephalography*): Electric potentials recorded on the human scalp; brain waves.

electric field: The spatial derivative (gradient) of electric potential. The force on a unit (standard) charge.

electroencephalography: See *EEG*.

electrophysiology. Studies of the electrical properties of brain tissue, including EEG, ECoG, LFP, and single-neuron recordings.

emergence. The process of creating new large-scale features through the actions of smaller entities that do not possess these features in isolation.

engram: The physical representation of memory.

entanglement: A fundamental holistic phenomenon in which groups of individual particles become correlated when confined to the same location. Individual particles cannot be described independently; rather, the system as a whole exists in a quantum state.

entropy: A system's inaccessible information, often about the microscopic particles forming a macroscopic system. A measure of disorder and ignorance.

epilepsy: A group of neurological disorders marked by recurrent sensory disturbances, loss of consciousness, and sometimes convulsions. Associated with abnormal EEG activity.

epistemology: The ways in which knowledge is acquired.

EPSP (*excitatory post-synaptic potential*): A membrane change in a target neuron induced by a neurotransmitter, thereby making it easier for the target neuron to fire its own action potential.

ERP (*event-related potential*): A late-appearing slow waveform resulting from a sensory stimulus, normally recorded on the scalp. A subcategory of EEG.

event-related potential: See ERP.

excitatory post-synaptic potential: See EPSP.

explanatory gap: The difficulty that materialistic theories have in explaining how physical properties can give rise to the mind.

falsifiability: A test to distinguish genuine science from non-science.

fiber tract: The pathway of a bundle of axons in the cortical white matter layer or in a nerve.

field: Almost any kind of dynamic pattern that varies over space and time, often with emphasis on its mathematical representation.

fMRI (*functional magnetic resonance imaging*): An experimental measure of blood-oxygen level in small tissue volumes (voxels).

fractal-like: Geometric shapes, physical objects, brain tissue, and so forth that exhibit fine structure at progressively smaller scales. Bona fide fractals require a mathematical property called self-similarity.

free will: Our ability to make our own (free) choices with minimal constraint by deterministic influences.

frequency: The rate of some oscillatory process repeated over time; expressed in cycles per second or Hertz (Hz)

frequency band: A range of frequencies, usually small. The alpha frequency band is defined as 8 to 13 Hz. Visible light occupies a very narrow frequency band of electromagnetic radiation.

functional connection: A correlation or predictive relationship of some kind between two systems.

functional localization: States in which subsystems act mostly independent of other subsystems.

functional magnetic resonance imaging: See *fMRI*.

fundamental frequency: The lowest natural frequency of a system.

gamma rhythm : Brain rhythms with frequencies greater than about 30 Hz.

general anesthesia: A medically induced coma.

global coherence: States in which subsystems combine to form a single integrated system, perhaps with emergent properties.

global workspace model: A conceptual framework that views consciousness as the global sharing of information between many local networks or "processors."

half sub-network: A sub-network employing exactly half the total number of nodes in the full system.

hard problem: The deep mystery of consciousness and self-awareness.

harmonics: Integer multiples of some fundamental frequency; a special case of overtones. Called "partials" by musicians.

Hertz: See *Hz*.

hormones: Chemical messengers that carry signals through the blood from the endocrine system to other cells in the body.

Hz (*Hertz*): The basic unit of oscillation frequency; the number of cycles per second.

inhibitory post-synaptic potential: See *IPSP*.

initial condition: In general use, the initial state of some dynamic system, that is, before it begins to evolve. In mathematical physics, the initial values of variables describing the system under study.

interference: The cancellation of the positive and negative parts of traveling waves when they intersect; often resulting in standing waves.

IPSP (*inhibitory post-synaptic potential*): A membrane change in a target neuron induced by a neurotransmitter, thereby making it more difficult for the target neuron to fire its own action potential.

link: A connection between network nodes.

lobe scale: Typical size of a cortical lobe, perhaps 10–15 cm.

locked-in syndrome: A medical condition in which the patient may be fully conscious but cannot move or communicate due to paralysis of voluntary muscles.

lucid dream: A state in which the dreamer is aware that he is dreaming.

macrocolumn: A cortical-tissue column defined by the local intra-cortical spread of individual pyramidal cells; contains about one million neurons within its diameter of approximately 3 mm.

magnetoencephalography: See *MEG*.

masking: An experimental procedure in which target images are flashed very briefly and conscious awareness is blocked by subsequent images.

materialism: The view that mental states are created by brain structure and function; that is, mind emerges from brain.

MCS (*minimally conscious state*): A brain disorder that preserves just a little bit of awareness.

measurement problem: In quantum mechanics, the interpretation of weird results. The problem of whether or not or how the wavefunction "collapses" such that consistent experimental results are obtained.

MEG (*magnetoencephalography*): Superconducting quantum technology used to record the brain's tiny magnetic fields.

metaphor: A system that has some properties that remind us of a different system, a weaker relationship than an analogue.

millisecond: One one-thousandth (0.001) of a second.

minicolumn: A cortical-tissue column defined by the spatial extent of intracortical inhibitory connections; contains about one hundred neurons within its diameter of approximately 0.03 mm.

minimally conscious state: See *MCS*.

modulation: The process of modifying one or more properties of some rhythm.

mokita: That which we all know to be true, but agree not to talk about.

The human tendency of hiding inconvenient truths under some pro-verbial rug. The elephant in the room.

monism: The position that mental or proto-mental properties occur at the fundamental level of physical reality.

motor cortex: The region of the cerebral cortex involved in the planning, control, and execution of voluntary movements.

multiscale conjecture: The view that consciousness is encoded at multiple scales and depends fundamentally on cross-scale interactions (circular causality). In this view, no special C-scale needs to exist.

multiverse: The hypothetical set of possible universes that comprise everything that exists.

myelination of axon: The process by which special white matter cells wrap around an axon and increase the propagation speed of the axon action potential.

natural frequency. A frequency at which a system tends to respond most strongly to external influences.

neuromodulator: Neurotransmitters that regulate widely dispersed neural populations.

neuron: Nerve cells that transmit signals to different parts of the brain and body.

neurotransmitter: Chemical that alters properties of specific target neurons.

node: An intersection where links come together in a network.

non-local interaction: Interactions that occur when distant parts of a system are connected without directly influencing locations in between. Not the same as *nonlocal interactions*.

nonlocal interaction: Quantum influence between distant events acting at faster-than-light speed. Not the same as *non-local interaction*.

ontology: The ultimate nature of reality.

oscillation. The up-and-down (or back-and-forth) movement of something, a sound wave, a playground swing, an EEG, and so forth.

overtone: Natural frequency higher than the fundamental frequency; may or may not be harmonic. Violin strings have harmonic overtones; the violin body has non-harmonic overtones.

P300: A slow waveform of the ERP appearing about three hundred milliseconds after a visual or auditory stimulus, normally recorded on the scalp.

perception: The awareness of something through the senses.

phase: The location of a sine wave along the time axis. Or, the time difference between the two peaks of two sine waves.

philosophical zombie: See *zombie*.

physicalism: See *materialism*.

pitch: Our conscious sensation of sound frequency.

Planck scale: The limiting scale (size) of space-time itself; a fundamental voxel size.

pre-conscious: Unconscious activity that directly influences conscious behavior.

prefrontal cortex: The front part of the frontal lobe of the cerebral cortex.

primary evoked potentials: Electrical responses to external stimuli recorded on the scalp near the primary sensory cortex.

principle of organizational invariance. The idea that any two systems with the same functional organization must exhibit qualitatively identical behavior and experience.

pyramidal cell: The most populous cortical neuron. Shaped like a tall tree with many branches (dendrites) and an extensive "root system" (axons).

p-zombie: See *zombie*.

quantum mechanics: The branch of physics that deals with phenomena at the atomic and subatomic scales.

rapid eye movement sleep: See *REM sleep*.

receptor: A molecular "lock" that accepts only specific molecular "keys."

regional scale: A relatively large level of organization, but less than the global scale. Similar to the lobe scale.

REM sleep (*rapid eye movement sleep*): The sleep stage most associated with dreaming, in which rapid eye movement may be observed under closed eyelids.

resonance: The tendency of a system to respond mainly to a narrow range of input frequencies.

reuptake: The reabsorption of a neurotransmitter into the pre-synaptic neuron.

rhythm: A repeated pattern of some signal or sound, like the oscillations of electric voltages in the brain.

RQTC conjecture: A postulated fundamental link between the physical and mental worlds, involving unknown connections of consciousness to relativity, quantum mechanics, and thermodynamics.

scale chauvinism: A bias shown by scientists in favor of experiments carried out at some favored level of organization or special scale.

schizophrenia: An illness causing severe mental disturbances. *Positive symptoms* may include delusions, thought disorders, and hallucinations. Schizophrenics may hear voices or believe people are reading their minds, controlling their thoughts, or plotting against them. *Negative symptoms* include a general lack of motivation or desire to form relationships, and blunted emotion.

selective attention: The process in which experimental subjects are instructed to pay attention to certain things and ignore everything else.

signal medium: A substance through which information is exchanged.

signatures of consciousness: In general use, consciousness correlates; that is, any brain measurement reliably correlated with mental activity. In more-specific use, brain measurements that clearly distinguish conscious from unconscious experiences.

sine wave: The smooth up-and-down movement of a waveform oscillating at some fixed frequency.

sinusoidal oscillations: See *sine wave*.

small-world network: A social, physical, brain, or other network in which a few long-distance (non-local) connections cause a substantial reduction in the path lengths between pairs of network nodes.

somatosensory cortex: A cortical region reacting to sensory input from the skin, including touch, pain, and temperature stimuli.

spatial resolution: The accuracy with which one can distinguish two or more closely spaced sources of activity.

spatial scales: Levels of organization.

spectral analysis: Estimates of individual frequency contributions to a signal; usually obtained by the computational methods of Fourier transforms.

split-brain patients: Epilepsy patients with their cortical hemispheres surgically separated.

spontaneous EEG: Ongoing brain rhythms that occur with no need of external sensory input.

SSVEP (*steady state visually evoked potentials*): Scalp potentials in the same narrow frequency band as a flickering-light stimulus; recorded while the subject is performing some task. A subcategory of *EEG*.

standing waves: Fixed spatial patterns that occur when multiple traveling waves come together to produce wave interference patterns. The underlying cause of many, if not most, resonance phenomena.

steady state visually evoked potentials: See *SSVEP*.

subconscious: See *pre-conscious*.

subliminal: Images presented below the threshold of conscious awareness, as in masking experiments.

synapse: The end of an axon where neurotransmitters are stored and released.

synaptic action field: The numbers of active excitatory or inhibitory synapses in some large tissue mass at each "instant" in time (actually time averaged over a very short interval).

synchronization: The condition of any two signals when they line up in time; that is, exhibit a zero or a very small phase difference. All synchronized signals are coherent, but the reverse need not be true.

theta rhythm: EEG oscillations with frequencies in the approximate range of 4–8 Hz.

thought experiment: A mental process that considers the experimental consequences of some new idea.

time: One of the four coordinates of our space-time universe; a very familiar entity, but one that is quite mysterious when considered in depth. Consciousness seems to be intimately connected with time in some way, but unfortunately nobody understands time.

top-down interaction: The influence of large and emergent systems on smaller-scale systems.

Tourette's syndrome. An inherited disease characterized by sudden repetitive movements and involuntary sounds, sometimes containing obscene words or derogatory remarks.

traveling wave: Local, temporary changes in medium properties that propagate through the medium, transmitting information and/or energy from one location to another. When traveling waves combine in confined spaces like violins or brains, standing waves occur.

Turing test: A thought experiment aimed to find out if a computer exhibits human-level intelligence.

ultra-information: That which distinguishes one thing from another. An extension of ordinary information to include things that are fundamentally hidden from direct observation but still influence the physical world in some way. Einstein's spooky action, Bohm's implicate order, Heisenberg's wavefunction collapse, parallel universes, and perhaps even consciousness—these seem to qualify for membership in this "club."

uncertainty principle: Places fundamental limits on the precision with which certain pairs of physical properties of a particle can be known simultaneously. Such pairs are known as *complementary variables*. Position and momentum (or velocity) is an example.

unconscious: Brain activity that lacks conscious awareness but may rise to the level of the pre-conscious.

vagus nerve stimulation: Electrical stimulation of one of the cranial nerves, providing indirect brain-input currents meant to prevent seizures.

vegetative state: See *VS*.

visual field: Everything one sees on either the right or the left side of his or her nose.

volume conduction: The passive spread of electric currents and potentials through tissue or other conductive medium.

voxel: a small-volume element analogous to a pixel, which represents a small surface area as in the example of a photograph.

VS (*vegetative state*): This unconscious (vegetative) state is characterized by a lack of response to loved ones or any other sensory stimuli.

Wernicke's area: A major language-understanding region located near the intersection of the temporal and parietal lobes.

white matter: The brain layer just below the cerebral cortex, consisting mostly of axons connecting different cortical locations to each other.

zombie: A creature or machine that behaves like a conscious human in every possible test but actually lacks consciousness. The critical C-factor is missing.

NOTES

A Note on the Citation Strategy Adopted Here

*T*he *New Science of Consciousness* borrows many ideas from at least five major fields of study—complexity science, physics, cognitive science, neuroscience, and philosophy. The number of potential citations to the many subtopics of these major fields is enormous, but probably counterproductive to my goal of serving a general audience. With this goal in mind, I have adopted the following citation strategy: Any idea or experimental outcome that is widely accepted within a major field is typically not referenced here, even though the idea may be unfamiliar to persons outside of that particular major field. Such information is, in any case, easily accessed on the Internet, from Google Scholar, *Wikipedia*, and many other sources. In this manner, my chosen citations are focused more on controversial or subtle ideas and information that may be more difficult to find, in addition to quoted material, of course. Furthermore, I have made only minimal attempts to provide proper credit for the major ideas. Most of my information stems from multiple sources that are difficult, if not impossible, to rank fairly. In particular, very few of the so-called "original" ideas originate with me.

CHAPTER 1: INTRODUCTION TO MIND AND BRAIN

1. Susan Blackmore, *Conversations on Consciousness* (New York: Oxford University Press, 2006).

2. The phrase "remembered present" is credited to Gerald Edelman. Our treatment of brains as genuine complex systems fits naturally with Edelman's *theory of neuronal group selection*, also known as *neural Darwinism*, which is based on three tenets—developmental selection, experiential selection, and reentry.

Developmental selection. While brain anatomy is controlled by genetic factors, enormous variability in the neural circuitry occurs between individuals. The functional plasticity of neural interconnections enables neuronal groups to self-organize into many adaptable networks or modules.

Experiential selection. A continuous process of synaptic selection occurs among neural networks during an individual's life. This selective process is analogous to the selection processes that act on populations of individuals in species; hence the popular tag, "neural Darwinism."

Reentry. The ongoing interchange of signals between brain networks and their dynamic information patterns. The reentry process continuously interrelates these patterns to each other in time and space—a form of higher-order selection that appears unique to animal brains. The cortico-cortical axons may provide the dominant reentry process in humans (see note 4 below). Gerald M. Edelman, *Bright Air, Brilliant Fire* (New York: Basic Books, 1992); Gerald M. Edelman and Giulio Tononi, *A Universe of Consciousness* (New York: Basic Books, 2000).

3. Christof Koch, *Consciousness. Confessions of a Romantic Reductionist* (Cambridge MA: MIT Press, 2012). A highly recommended personal account of Koch's well-known studies of consciousness, including Koch's close collaboration with Francis Crick over many years.

4. *On the subject of brain waves:* Physicists lacking neuroscience knowledge may question the origin of genuine *brain waves* because such low-frequency waves cannot be "electromagnetic" in small systems like brains. Electromagnetic waves generally require high-frequency coupling between electric and magnetic fields—the usual magnetic induction described by Maxwell's equations. In contrast to electromagnetic waves, the proposed brain waves are based on the finite speeds of action potential propagation in the cortico-cortical axons and the length distribution of these axons. Cortico-cortical axon speeds are distributed but peak in the 5 to 10 meter/second range; axon lengths range up to about 15 centimeters. The corresponding global delays are roughly 15 to 30 milliseconds, implying cortical/white matter natural frequencies in the 5 to 10 Hz range and above. Together with the local delays due to neuron post synaptic potential (PSP) rise and decay times, the global delays provide plausible bases for the cortex's natural frequencies. These waves of *cortical synaptic action* are believed to generate several EEG phenomena, including a global alpha rhythm, as suggested in a dozen or so papers published since 1972. For physics aficionados—the basic model involves integral and differential equations as outlined in a recent review: Paul L. Nunez and Ramesh Srinivasan, "Neocortical Dynamics Due to Axon Propagation Delays in Cortico-Cortical Fibers: EEG Traveling and Standing Waves with Implications for Top-Down Influences on Local Networks and White Matter Disease," *Brain Research* 1542 (2013): 138–66.

5. My first contact with Gerald Edelman involved a fiasco similar to my first meeting with Crick. I was invited to discuss my research with several scientists at the Neurosciences Institute in La Jolla where Edelman was the director. "Why don't you drop by around 2:00," one said. I arrived at about 2:20 to find an auditorium half filled with impatient scientists, including Edelman, who had been waiting for twenty minutes for my "lecture" to begin. What lecture? It was news to me. At least this time, no wine was involved. Fortunately Edelman, who was famous for not suffering fools lightly, was friendly about the misunderstanding.

CHAPTER 2: THE SCIENCE AND PHILOSOPHY OF MIND

1. Pierre-Simon Laplace, *A Philosophical Essay on Probabilities*, trans. F. W. Truscott and F. L. Emory (New York: Dover, 1951), p. 4. Translated into English from the original French 6th edition.

2. Francis Crick, *The Astonishing Hypothesis: The Scientific Search for the Soul*, repr. ed. (New York: Scribner, 1995).

3. Wikiquote, "Wolfgang Pauli," https://en.wikiquote.org/wiki/Wolfgang_Pauli (accessed August 1, 2016); also cited in Oliver Burkeman, "Briefing: Not Even Wrong," *Guardian*, September 19, 2005, https://www.theguardian.com/science/2005/sep/19/ideas. g2 (accessed August 1, 2016).

4. Some atheists have suggested that agnostics are essentially "atheists without balls," a sentiment that I don't share—thus my adoption of the label "unequivocal agnostic." It seems to me that one can plausibly conjecture the existence of progressively more advanced intelligent entities in our universe, or even in the putative multiverse. At what advanced level these "creatures" might be considered gods or even God-like depends strongly on our definition of the emotionally loaded "G-word." Still another agnostic argument employs modern physics, which says that some things are far more mysterious than simply unknown; rather they are *fundamentally unknowable*, as discussed in chapter 10 in the context of ultra-information. Thus, one can advance a plausible argument that the agnostic position is the most "scientific" religious-philosophical position.

5. John Locke, *An Essay Concerning Human Understanding*, bk. 4, *Of Knowledge and Probability*, chap. 10, "Of Our Knowledge of the Existence of a God" (London, 1690), available at Enlightenment, http://enlightenment.supersaturated.com/johnlocke/BOOKIVChapterX.html (accessed August 1, 2016).

6. David J. Chalmers, *The Character of Consciousness* (New York: Oxford University Press, 2010). John R. Searle, *The Rediscovery of the Mind* (Cambridge, MA: MIT Press, 1992). Daniel C. Dennett, *Consciousness Explained* (Boston: Little, Brown, 1991).

7. Edward N. Zalta, ed., "Donald Davidson," *Stanford Encyclopedia of Philosophy*, May 29, 1996, http://plato.stanford.edu/archives/win2012/entries/davidson/.

8. Daniel Dennett, quoted in *Historical Dictionary of Quotations in Cognitive Science: A Treasury of Quotations in Psychology, Philosophy, and Artificial Intelligence*, comp. Morton Wagman (Westport, CT: Greenwood, 2000), p. 50.

9. Private communication from Richard Feynman to my colleague Lester Ingber.

10. Valentino Braitenberg, *Vehicles: Experiments in Synthetic Psychology* (Cambridge, MA: MIT Press, 1984).

11. James Hogan, *Code of the Lifemaker* (New York: Ballantine Books, 1983).

12. Michael S. Gazzaniga, *Tales from Both Sides of the Brain* (New York: Harper Collins, 2015).

13. Benjamin Libet, *Mind Time* (Cambridge, MA: Harvard University Press, 2004).

CHAPTER 3: A BRIEF LOOK INTO BRAIN STRUCTURE AND FUNCTION

1. Hermann Haken, *Synergetics: An Introduction*, 3rd ed. (Berlin: Springer-Verlag, 1983); Karl J. Friston, Guilio Tononi, Olaf Sporns, and Gerald. M. Edelman, "Characterizing the Complexity of Neuronal Interactions," *Human Brain Mapping* 3 (1995): 302–14; Neil Johnson, *Simple Complexity* (Oxford, UK: Oneworld Publications, 2007); Misha Z. Pesenson, *Multiscale Analysis and Nonlinear Dynamics: From Genes to Brain* (Weinheim, Germany: Wiley-VCH, 2012). Hermann Haken developed the field of synergetics, which emphasizes the importance of circular causality in complex systems.

2. A discussion of the possible roles of neurotransmitters acting selectively at different cortical depths to move cortical dynamic behavior between more localized (hypocoupled) to more globally coherent (hypercoupled) states is provided by Richard B. Silberstein, "Neuromodulation of Neocortical Dynamics," in Paul L. Nunez, *Neocortical Dynamics and Human EEG Rhythms* (New York: Oxford University Press, 1995), pp. 591–627.

The following books advance strong arguments about brain complexity and that brain complexity and consciousness are strongest at intermediate brain states between local and global:

Gerald M. Edelman, *Bright Air, Brilliant Fire* (New York: Basic Books, 1992).
Gerald M. Edelman and Giulio Tononi, *A Universe of Consciousness: How Matter Becomes Imagination* (New York: Basic Books, 2000).

3. Ibid.
4. Ibid.

5. Valentino Braitenberg and Almut Schuz, *Anatomy of the Cortex: Statistics and Geometry* (New York: Springer-Verlag, 1991); Janos Szentagothai, "Local Neuron Circuits of the Neocortex," in F. O. Schmitt and F. G. Worden, eds., *The Neurosciences 4th Study Program* (Cambridge, MA: MIT Press, 1979), 399–415.

The anatomist Valentino Braitenberg has long emphasized an important quantitative difference in the brains of different species of mammals. Following up on this idea, some years ago my colleague Ron Katznelson consulted the anatomical literature to estimate *Ratio*, the ratio of cortico-cortical to thalamo-cortical axons entering and leaving the underside of the cortex, as shown in figure 3-12. This figure also appears in his doctoral dissertation and his chapter in my book, "Normal Modes of the Brain: Neuroanatomical Basis and a Physiological Theoretical Model," in Paul L. Nunez, *Electric Fields of the Brain: The Neurophysics of EEG*, 1st ed. (New York: Oxford University Press, 1981), pp. 401–42.

6. Lester Ingber's numerous papers on the statistical mechanics of neocortical interactions as well as applications in other fields where cross-scale interactions are essential features may be found at http://www.ingber.com/. Here are two examples:

Lester Ingber, "Statistical Mechanics of Neocortical Interactions. I. Basic Formulation," *Physica D*5 (1982): 83–107.

Lester Ingber, "Statistical Mechanics of Neocortical Interactions: Path-Integral Evolution of Short-Term Memory," *Physical Review E* 49 (1994): 4652–64, http://www.ingber.com/smni94_stm.pdf.

The central importance of the brain's nested hierarchy of structure is considered from a medical perspective in the following:

Todd E. Feinberg, *From Axons to Identity: Neurological Explorations of the Nature of the Self* (New York: Norton, 2007).

Todd E. Feinberg, "Neuroontology, Neurobiological Naturalism, and Consciousness: A Challenge to Scientific Reduction and a Solution," *Physics of Life Reviews* 9 (2012): 13–34.

Todd E. Feinberg and Jon M. Mallatt, *The Ancient Origins of Consciousness: How the Brain Created Experience* (Cambridge, MA: MIT Press, 2016).

7. The confluence of physics and brain anatomy/physiology is discussed in the following book that includes six chapters by guest authors from different backgrounds:

Paul L. Nunez, *Neocortical Dynamics and Human EEG Rhythms* (New York: Oxford University Press, 1995). See also the references in notes 1 and 5 of this chapter.

8. See the references in note 2 of this chapter.

9. See the references in note 5 of this chapter.

10. See the references in note 2 of this chapter.

11. Wendell J. S. Krieg, *Connections of the Cerebral Cortex* (Evanston, IL: Brain Books, 1963). Wendell J. S. Krieg, *Architectonics of Human Cerebral Fiber System* (Evanston, IL: Brain Books, 1973). See also the references in note 5 of this chapter.

12. See the references in notes 2 and 5 of this chapter.

13. Antonio Damasio, *Self Comes to Mind: Constructing the Conscious Brain* (New York: Pantheon Books, 2010).

14. Charles R. Darwin, *The Expression of the Emotions in Man and Animals* (London: J. Murray, 1872).

15. See reference to Ron Katznelson in note 5.

CHAPTER 4: STATES OF MIND

1. Stuart Sutherland, ed., *Macmillan Dictionary of Psychology* (London: Palgrave Macmillan, 1989).

2. Thomas Nagel, "What Is It Like to Be a Bat?" *Philosophical Review* 83, no. 4 (1974): 435–50, http://organizations.utep.edu/portals/1475/nagel_bat.pdf.

3. Olaf Sporns, *Networks of the Brain* (Cambridge, MA: MIT Press, 2011).

4. Figure 4-2 was inspired by a similar figure in the chapter: Nicholas D. Schiff, "Large-Scale Brain Dynamics and Connectivity in the Minimally Conscious State," in *Handbook of Brain Connectivity*, edited by Viktor K. Jirsa and Anthony R. McIntosh (Berlin: Springer-Verlag, 2007): 505–20.

5. Much of the material on mental states in chapters 4 and 9 relies on the following comprehensive edited text with many distinguished chapter authors: Steven Laureys and Giulio Tononi, eds., *The Neurology of Consciousness* (Amsterdam: Academic Press, 2009). In addressing the hard problem, Tononi has developed the interesting idea of integrated information, defined as the information generated by interacting elements over and above information generated by the same isolated elements. See chapter 8.

6. My former student Brett M. Wingeier helped develop the RNS system for seizure detection and "shock stoppage" while working at NeuroPace, Inc. Figures 6-10, 6-14, 6-15, and 7-7 and the associated analyses are based on his doctoral research.

7. Michael Crichton, *The Terminal Man* (1972; rpr. New York: Avon, 2002).

8. By signaling only with his eyelid, Bauby dictated the entire book *The Diving Bell and the Butterfly* (New York: Free Press, 2006).

9. The field of brain-computer interfaces is nicely presented in the following edited text by multiple authors from different backgrounds: Jonathan R. Wolpaw and Elizabeth

W. Wolpaw, eds., *Brain-Computer Interfaces for Communication and Control* (New York: Oxford University Press, 2012).

CHAPTER 5: SIGNATURES OF CONSCIOUSNESS

1. Paul Davies, *About Time* (New York: Simon & Schuster, 1995); Stephen Hawking, *A Brief History of Time* (New York: Bantam Books, 1998). Roberto M. Unger and Lee Smolin, *The Singular Universe and the Reality of Time: A Proposal in Natural Philosophy* (Cambridge, UK: Cambridge University Press, 2014).

2. John Wheeler, quoted in Paul Davies, *About Time: Einstein's Unfinished Revolution*, 1st ed. (Simon & Schuster, 1996).

3. Benjamin Libet, *Mind Time* (Cambridge, MA: Harvard University Press, 2004); Antonio Damasio, *Self Comes to Mind: Constructing the Conscious Brain* (New York: Pantheon Books, 2010).

4. Bernard Baars, *A Cognitive Theory of Consciousness* (Cambridge, UK: Cambridge University Press, 1989).

5. David Hubel and Torsten Wiesel were co-recipients of the 1981 Nobel Prize in Physiology or Medicine (shared with Roger Sperry) as a result of their work on the visual system. See note 5 of chapter 8.

6. Stanislas Dehaene, *Consciousness and the Brain* (New York: Penguin Books, 2014).

7. Ibid.

8. Yogi Berra, *The Yogi Book: I Really Didn't Say Everything I Said!* (New York: Workman Publishing, 1998), p. 48.

9. Dehaene, *Consciousness and the Brain*.

10. Steven Laureys and Giulio Tononi, eds., *The Neurology of Consciousness* (Amsterdam: Academic Press, 2009).

11. Ibid.

12. Olaf Sporns, *Networks of the Brain* (Cambridge, MA: MIT Press, 2011).

13. R. Douglas Fields, "White Matter in Learning, Cognition and Psychiatric Disorders," *Trends in Neuroscience* 31 (July 2008): 361–70; R. Douglas Fields, "White Matter Matters," *Scientific American* 298 (2008): 54–61.

14. Paul L. Nunez and Ramesh Srinivasan, *Electric Fields of the Brain: The Neurophysics of EEG*, 2nd ed. (New York: Oxford University Press, 2006); Paul L. Nunez, *Brain, Mind, and the Structure of Reality* (New York: Oxford University Press, 2010); Paul L. Nunez and Ramesh Srinivasan, "Neocortical Dynamics Due to Axon Propagation Delays in Cortico-Cortical Fibers: EEG Traveling and Standing Waves with Implications for Top-Down Influences on Local Networks and White Matter Disease,"

Brain Research 1542 (2014): 138–66; Paul L. Nunez, Ramesh Srinivasan, and R. Douglas Fields, "EEG Functional Connectivity, Axon Delays and White Matter Disease," *Clinical Neurophysiology* 126 (2015): 110–20.

    15. Sporns, *Networks of the Brain*.

    16. R. Douglas Fields, see notes 13 and 14.

CHAPTER 6: RHYTHMS OF THE BRAIN

    1. The title of chapter 6 is borrowed from the following text, which contains broad discussions of both animal and human EEG experiments: György Buzsáki, *Rhythms of the Brain* (New York: Oxford University Press, 2006).

Extensive classical and contemporary studies of the alpha rhythm are reviewed in John C. Shaw, *The Brain's Alpha Rhythm and the Mind* (Amsterdam: Elsevier, 2003).

    2. Some neuroscientists lacking physics backgrounds are confused about the fundamentals of electromagnetic fields. This is not meant as a criticism—EEG and cognitive science are highly interdisciplinary, and no one scientist can be an expert in more than a few fields of knowledge. Here are a few common misconceptions:

> Magnetoencephalography (MEG), which records the tiny magnetic fields generated by the brain, was developed around 1980. It has often been claimed that MEG is more accurate than EEG, but this claim is mostly false. One publication may make this claim, and then another will parrot the same claim without supporting evidence. The short story is this—MEG holds one major advantage over EEG because tissue is transparent to magnetic fields. On the other hand, MEG suffers a major disadvantage because, thus far, its sensors must be placed a centimeter or two above the scalp, more than twice as far from cortical sources as EEG electrodes. Future technology may shift this balance, but in any case, MEG and EEG are selectively sensitive to different sets of cortical sources, thereby providing complementary information about brain patterns.
>
> At the low frequencies of interest in electrophysiology, the electric and magnetic fields are *uncoupled*; that is, each may be estimated without reference to the other. For this reason, one should avoid the provocative label *electromagnetic* in connection with EEG or MEG—this erroneous label implies a single (coupled) field, which generally exhibits more-complicated dynamic behavior than the quasi-static fields of EEG and MEG. If brains actually generated "electromagnetic fields" similar to those produced by cell phones, humans might be able to use this putative extrasensory process to communicate. But the brain's electric and magnetic fields fall off sharply within a few centimeters outside the head, essentially because the

fields are very low-frequency and uncoupled to each other; they are "electric and magnetic" rather than "electromagnetic." Suppose for purposes of argument that extrasensory perception (ESP) were proven to be real phenomenon. We would be pretty sure that any such ESP would be unrelated to electromagnetic fields; some entirely different process would have to be involved.

Paul L. Nunez and Ramesh Srinivasan, *Electric Fields of the Brain: The Neurophysics of EEG*, 2nd ed. (New York: Oxford University Press, 2006); Paul L. Nunez, "Electric and Magnetic Fields Produced by Brain Sources," in Jonathan R. Wolpaw and Elizabeth W. Wolpaw, eds., *Brain-Computer Interfaces for Communication and Control* (New York: Oxford University Press, 2012), pp. 45–63.

3. Chapters 2 and 3 of Paul L. Nunez, *Neocortical Dynamics and Human EEG Rhythms* (New York: Oxford University Press, 1995) contain discussions of standing waves in brains and musical instruments, including images of standing waves in violin bodies.

4. Ibid.

5. Mark Twain, requoting humorist Edgar Wilson "Bill" Nye, in *Mark Twain's Autobiography*, http://www.twainquotes.com/Opera.html (accessed August 1, 2016).

6. See note 3 of this chapter.

7. Herbert Jasper and Wilder Penfield, *Epilepsy and the Functional Anatomy of the Human Brain* (London: Little, Brown, 1954).

8. John S. Barlow, *The Electroencephalogram: Its Patterns and Origins* (Cambridge, MA: MIT Press, 1993); Erol Basar, Martin Schurmann, Canan Basar-Eroglu, and Sirel Karakas, "Alpha Oscillations in Brain Functioning: An Integrative Theory," *International Journal of Psychophysiology* 26 (1997): 5–29.

9. See Jasper and Penfield, *Epilepsy and the Functional Anatomy of the Human Brain*.

10. Donald L. Schomer and Fernando H. Lopes da Silva, eds., *Niedermeyer's Electro-encephalography: Basic Principles, Clinical Applications, and Related Fields*, 6th ed. (Philadelphia: Lippincott, Williams, and Wilkins, 2010); John S. Ebersole, ed., *Current Practice of Clinical Encephalography*, 4th ed. (Philadelphia: Wolters Kluwer Health, 2014).

11. Wolf Singer and Charles M. Gray, "Visual Feature Integration and the Temporal Correlation Hypothesis," *Annual Review of Neuroscience* 18 (1995): 555–86; Paul L. Nunez and Ramesh Srinivasan, "Scale and Frequency Chauvinism in Brain Dynamics: Too Much Emphasis on Gamma Band Oscillations," *Brain Structure and Function* 215 (2010): 67–71 (this was an invited "Brain Mythology" contribution, hence the purposely provocative title.); Buzsaki, *Rhythms of the Brain*.

12. Walter J. Freeman, *How Brains Make Up Their Minds* (New York: Columbia University Press, 2000); Gerald M. Edelman, *Bright Air, Brilliant Fire* (New York: Basic Books, 1992).

13. Fred Hoyle, *The Black Cloud* (New York: New American Library, 1957).

14. Paul L. Nunez and Ramesh Srinivasan, "Fallacies in EEG," in *Electric Fields of the Brain: The Neurophysics of EEG*, 2nd ed. (New York: Oxford University Press, 2006), pp. 56–98.

15. W. Grey Walter, "Normal Rhythms—Their Development, Distribution and Significance," in *Electroencephalography; A Symposium on Its Various Aspects*, edited by Denis Hill and Goeffery Parr (Oxford, England: Macdonald, 1950), pp. 203–27.

16. Nunez, *Neocortical Dynamics and Human EEG Rhythms*; Brett M. Wingeier, "A High Resolution Study of Coherence and Spatial Spectra in Human EEG" (PhD dissertation, Tulane University, 2004); Nunez and Srinivasan, *Electric Fields of the Brain*. In different subjects, correlations between mental calculations and increased amplitude may be stronger in either the theta or upper alpha band; also, the strength of reduced lower-band amplitudes may depend on task difficulty. Different subjects may employ different "dynamic strategies" to achieve the same goal.

CHAPTER 7: BRAIN SYNCHRONY, COHERENCE, AND RESONANCE

1. Electromagnetic waves that travel along power or communication lines rely on strong coupling between electric and magnetic fields (*magnetic induction*), otherwise, no wave propagation can occur in these lines. In sharp contrast, action potentials occur as a result of the (naturally) selective nonlinear behavior of neural membranes. For physics aficionados: this nonlinear diffusion process produces traveling waves along axons. While actual magnetic induction is entirely negligible in axon propagation, the membrane nonlinearity creates a phenomenon that has been characterized as an "effective magnetic induction," possibly adding to the confusion of some neuroscientists.

2. Very early on, Walter Freeman recognized the essential problem of scale inherent in neuroscience work at mesoscopic (intermediate) levels. Theoretical connections between the microscopic-level activity of neurons in small neural networks and the mesoscopic-scale activity of cell assemblies was (and still is) poorly understood. In the 1960s, the biophysicist Aharon Katzir-Katchalsky suggested treating cell assemblies using network thermodynamics. This idea was adopted by Freeman, who proposed the K-sets, defined as cell assemblies that form a nested hierarchy of multiscale networks. Freeman's numerous papers on multiscale issues, nonlinear brain dynamics, and more-philosophical topics may be found at http://sulcus.berkeley.edu. Here is a small sample:

Walter J. Freeman, *Mass Action in the Nervous System* (New York: Academic Press, 1975).

Walter J. Freeman and Michael Breakspear, "Scale-Free Neocortical Dynamics," *Scholarpedia* 2, no. 2 (2007): 1357, http://www.scholarpedia.org/article/Scale-free_neocortical_dynamics.

Walter J. Freeman and Harry Erwin, "Freeman K-sets," *Scholarpedia* 3, no. 2 (2008): 3238, http://www.scholarpedia.org/article/Freeman_K-sets. For a means of formal scale crossing using modern statistical mechanics, see the works of Lester Ingber cited in note 6 of chapter 3.

3. The relationship between small-scale synaptic sources in the cortical depths to the larger-scale cortical *mesosources* (current dipole moments per unit volume) is treated in detail in chapter 4 of Paul L. Nunez and Ramesh Srinivasan, *Electric Fields of the Brain: The Neurophysics of EEG*, 2nd ed. (New York: Oxford University Press, 2006).

4. Since 1972 I have promoted the idea that these (well-established) traveling waves of synaptic action often combine (interfere) to form standing waves, which then contribute substantially to several observed EEG phenomena; however, this issue remains somewhat controversial. In this context, the following publication includes my review article, attached commentaries by eighteen neuroscientists, and my replies to their commentaries: Paul L. Nunez, "Toward a Quantitative Description of Large-Scale Neocortical Dynamic Function and EEG," *Behavioral and Brain Sciences* 23 (2000): 371–437.

A later, updated review covering similar topics is Paul L. Nunez and Ramesh Srinivasan, "Neocortical Dynamics Due to Axon Propagation Delays in Cortico-Cortical Fibers: EEG Traveling and Standing Waves with Implications for Top-Down Influences on Local Networks and White Matter Disease," *Brain Research* 1542 (2014):138–66.

5. Alan Gevins pioneered studies of the functional connections associated with mental activity (the so-called shadows of thought) using both low- and high-resolution EEG for more than forty years. His numerous papers may be found at http://independent.academia.edu/AlanGevins/Papers. Here is one example where high-resolution EEG was employed: Alan S. Gevins, Michael E. Smith, Linda McEvoy, and Daphne Yu, "High-Resolution Mapping of Cortical Activation Related to Working Memory: Effects of Task Difficulty, Type of Processing, and Practice," *Cerebral Cortex* 7 (1997): 374–85.

6. We call the high-resolution algorithm used to construct figure 7-7 the *New Orleans spline Laplacian*. Several other Laplacian algorithms have been developed independently in other labs; they are expected to yield very similar results. The New Orleans algorithm has been verified with more than a thousand simulations employing different source distributions, head models, electrode placements, and noise levels. It has also been shown to agree closely with an alternate approach to high-resolution EEG called *Melbourne dura imaging*, which was developed at the Brain Science Institute (see note 10 below). High-resolution EEG is discussed in several contexts in chapters 2, 8, 9, and 10 of Nunez and Srinivasan, *Electric Fields of the Brain*.

7. Ibid.

8. Ibid.

9. My former student Ramesh Srinivasan is currently chairman of the Department of Cognitive Science at the University of California at Irvine and operates the Human

Neuroscience Lab (http://hnl.ss.uci.edu/), where numerous papers are listed. Here are three example publications that discuss frequency tagging:

Ramesh Srinivasan, D. Patrick Russell, Gerald M. Edelman, and Guilio Tononi, "Frequency Tagging Competing Stimuli in Binocular Rivalry Reveals Increased Synchronization of Neuromagnetic Responses during Conscious Perception," *Journal of Neuroscience* 19 (1999): 5435–48.

Ramesh Srinivasan, William R. Winter, Jian Ding, and Paul L. Nunez, "EEG and MEG Coherence: Measures of Functional Connectivity at Distinct Spatial Scales of Neocortical Dynamics," *Journal of Neuroscience Methods* 166 (2007): 41–52.

Gerald M. Edelman and Giulio Tononi, *A Universe of Consciousness* (New York: Basic Books, 2000).

10. Neuro-Insight, Inc., is a neuroscience-based market-research company founded by Richard Silberstein, who continues as CEO. He was also founder and director of the Brain Science Institute at Swinburne University in Melbourne, Australia, from 1996 to 2002, where he developed a number of different scientific applications of steady-state visually evoked potentials (SSVEP), essentially the same technology as frequency tagging. *Melbourne dura imaging* was developed at this institute by Peter Cadusch and Silberstein. (Disclaimer: I have a small financial interest in Neuro-Insight.) Some of the scientific basis for this technology may be found at http://www.neuro-insight.com/scientific-research/. Here is one paper that employs frequency tagging: Richard B. Silberstein, Frank Danieli, and Paul L. Nunez, "Fronto-Parietal Evoked Potential Synchronization Is Increased during Mental Rotation," *NeuroReport* 14 (2003): 67–71.

11. Ibid.

12. Ibid.

13. See references in note 9 of this chapter.

CHAPTER 8: NETWORKS OF THE BRAIN

1. Danielle S. Bassett and Edward T. Bullmore, "Human Brain Networks in Health and Disease," *Current Opinion in Neurobiology* 22 (2009): 340–47; Danielle S. Bassett and Michael S. Gazzaniga, "Understanding Complexity in the Human Brain," *Trends in Cognitive Science* 15 (2011): 200–209; Olaf Sporns, *Networks of the Brain* (Cambridge, MA: MIT Press, 2011); Sebastian Seung, *Connectome: How the Brain's Wiring Makes Us Who We Are* (New York: Mariner Books, 2013). While this latter book is recommended, its title is not. My interpretation of current neuroscience indicates that "who we are" depends on far more than just the brain's wiring diagram. I suspect that the so-called connectome is only a very small, if important, part of the big picture of consciousness.

2. Resonance in discrete and continuous mechanical and electrical systems is

discussed in many elementary physics and engineering books. The case of nearest-neighbor plus next nearest-neighbor and so on up to thirty-two nearest-neighbor interactions in a loaded string, plus other physical analogues of brain resonance are treated in the following book: Paul L. Nunez, *Neocortical Dynamics and Human EEG Rhythms* (New York: Oxford University Press, 1995).

3. Ibid.

4. Since the publication of the following works based on networks and *graph theory*, the scientific literature has experienced an explosion of theoretical and experimental network studies in many disparate fields: Duncan J. Watts and Steven H. Strogatz, "Collective Dynamics of "Small-World" Networks," *Nature* 393 (1998): 440–42; Duncan J. Watts, *Small Worlds, The Dynamics of Networks Between Order and Randomness* (Princeton, NJ: Princeton University Press, 1999); Albert-Laszlo Barabasi, *Linked. The New Science of Networks* (Cambridge, MA: Perseus, 2002).

5. Scientific accounts of visual perception include discussion of the classic work by David Hubel and Torsten Wiesel. These neuroscientists shared the 1981 Nobel Prize with Roger Sperry, who was honored for his studies of split-brain patients. The Hubel/Wiesel experiments greatly expanded scientific knowledge of visual processing through a partnership that lasted over twenty years. Hubel and Wiesel showed how the visual system constructs complex representations of visual information from simple stimulus features. Visual signals are processed by the brain to generate edge detectors, motion detectors, depth detectors, and color detectors, the basic building blocks of the visual scene. In contrast, our current understanding of our perception of complex images like faces is quite limited.

6. The number of ways to choose a subset of k nodes from a set of n nodes is given by the binomial coefficient

$$\frac{n!}{k!(n-k)!}$$

where $n!$ is read as "n factorial." For example, 4! equals (4)(3)(2)(1), or 24. With $n = 10$ nodes and $k = 5$ nodes, the number of possible half sub-networks is

$$\frac{10!}{5!(10-5)!} = 252.$$

7. Mathematicians Frank Hoppensteadt and Eugene Izhikevich (HI) studied a mathematical model consisting of an arbitrary number (N) *of semi-autonomous oscillators*, assumed to be *pair-wise weakly connected* to themselves and to a "central" oscillator, represented by X_0, which may act to control the other oscillators. What exactly do HI mean by the label "oscillator"? Our interest in their work arises from the fact that a broad class of systems, including many kinds of neural network models, fit their oscillator category. That is, we could replace the label "oscillator" with many similar tags—*network, vibrator, resonator, rhythm creator*, and so on.

Oscillators (networks), labeled X_1, X_2, and so forth, are assumed to be pairwise weakly connected to themselves and to a so-called "central" oscillator X_0. While HI imagined X_0 to be a local network in the thalamus, I have suggested that X_0 might just as easily represent a global field or global network that acts top-down on local networks. The boldface symbol $\mathbf{X}_n(t)$ labels the dependent variables (unknowns) in the distinct networks (X_1, X_2, and so forth). For clarity, I have used the same symbols for mathematical functions as are used for the systems shown in figure 8-7. The entire system of N local oscillators (networks) is described by the vector of dependent variables $\mathbf{X}_n(t) = \{x_{n1}(t), x_{n2}(t), \ldots, x_{nK}(t)\}$, which indicates $K(n)$ scalar dependent variables (unknowns) for each weakly coupled network n. The coupled equations are assumed to take the general form

$$\frac{d\mathbf{X}_n}{dt} = \mathbf{F}_n(\mathbf{X}_n) + \varepsilon \sum_{j=1}^{N} \mathbf{G}_{nj}(\mathbf{X}_n, \mathbf{X}_j, \mathbf{X}_0, \varepsilon) \quad n = 1, N.$$

Here ε is the coupling constant between the networks; ε is small since the networks are "semi-autonomous." The vector functions $\mathbf{F}_n$ and $\mathbf{G}_{nj}$ are largely arbitrary. The main point is that the local networks only "substantially interact" (interact on a timescale of $1/\varepsilon$) when certain resonant relations exist between the local, semi-autonomous networks.

These networks might represent the dynamic activity of any one of many kinds of systems—mechanical, electrical, chemical, global synaptic action, or neural network—nearly any system governed by a set of differential equations producing non-chaotic dynamics. When (mathematically) disconnected from the larger system, each network X_n is assumed to produce non-chaotic (*quasi-periodic*) dynamics with a set of natural (resonant) frequencies ($f_{n1}, f_{n2}, f_{n3}, \ldots$). If the governing oscillator equations are *strongly connected*, we expect strong dynamic interactions and new resonant frequencies to emerge in the larger system of coupled oscillators, but we are more interested here in the case of *weak connections*. Typically, application of neural network models to genuine brain tissue is severely limited because of model dependence on unknown parameters, often with ambiguous or nonexistent physiological interpretations. The strength of this resonance theory lies in its independence from the unsupported assumptions typical of many neural-network models.

HI showed that the individual, weakly connected oscillators interact strongly only if certain resonant relations exist between the characteristic frequencies of the autonomous (unconnected) oscillators. Their oscillator category fits a large class of systems described by differential equations, linear or nonlinear. To take a simple example, suppose X_0 represents a global field of synaptic action with a single (autonomous) resonant frequency f_{01}, and this global oscillator is weakly connected to two local oscillators (X_1, X_2) also weakly connected to each other, as indicated in figure 8-7. Remember that the first frequency subscript, 0, indicates the global system X_0, and the second subscript, 1, indicates the fundamental frequency of that system. X_1 and X_2 might represent local cortical-thalamic networks, as long as the conditions of semi-autonomy and non-chaotic oscillations are satisfied. Assume that each of the two local networks has only a single natural frequency (f_{11} and f_{21}). In this example, the cortical-thalamic networks substantially interact only when the three natural frequencies (f_{01}, f_{11}, f_{21}) satisfy the resonant relation

$$m_0 f_{01} + m_1 f_{11} + m_2 f_{21} = 0.$$

Here (m_1, m_2) are any combination of non-zero integers and m_0 is any integer including zero. Two simple cases are $f_{11} = f_{21}$ and $f_{01} = f_{11} - f_{21}$.

The first case is the well-known resonant interaction between two oscillators with equal resonant frequencies, but many other resonant interactions are predicted for various combinations of the m indices. Suppose two thalamo-cortical networks have the distinct gamma resonant frequencies $f_{21} = 37$ Hz and $f_{11} = 47$ Hz. The two networks may substantially interact, perhaps forming a temporary (functionally connected) single network, when the global resonant frequency satisfies one of the resonant relations. For example, a presence of a global 10 Hz alpha rhythm is predicted to cause the local gamma networks to interact more strongly. The following publications discuss these issues:

Frank C. Hoppensteadt and Eugene M. Izhikevich, "Thalamo-Cortical Interactions Modeled by Weakly Connected Oscillators: Could Brains Use FM Radio Principles?" *Biosystems* 48 (1998): 85–92.

Eugene M. Izhikevich, "Weakly Connected Quasi-Periodic Oscillators, FM Interactions, and Multiplexing in the Brain," *SIAM Journal of Applied Mathematics* 59 (1999): 2193–2223.

Paul L. Nunez, *Brain, Mind, and the Structure of Reality* (New York: Oxford University Press, 2010).

Ramesh Srinivasan, Samuel Thorpe, and Paul L. Nunez, "Top-Down Influences on Local Networks: Basic Theory with Experimental Implications," *Frontiers in Computational Neuroscience* 7 (2013): 29.

Misha Z. Pesenson, ed., *Multiscale Analysis and Nonlinear Dynamics* (New York: Wiley, 2013).

Paul L. Nunez and Ramesh Srinivasan, "Neocortical Dynamics Due to Axon Propagation Delays in Cortico-Cortical Fibers: EEG Traveling and Standing Waves with Implications for Top-Down Influences on Local Networks and White Matter Disease," *Brain Research* 1542 (2014):138–66.

8. Ibid.

9. Ibid.

10. Ibid.

11. Cross-frequency coupling of brain rhythms in working memory, long-term

memory storage, visual perception, and other brain operations has been a hot topic in neuroscience for at least a decade or so. The following papers provide just a few examples:

Tamer Demiralp, Zubeyir Bayraktaroglu, Daniel Lenz, Stefanie Junge, Niko A. Busch, Burkhard Maess, Mehmat Ergen, and Christoph S. Herrmann, "Gamma Amplitudes Are Coupled to Theta Phase in Human EEG during Visual Perception," *International Journal of Psychophysiol*ogy 64 (2007): 24–30.

Nikolai Axmacher, Melanie M. Henseler, Ole Jensen, Ilona Weinreich, Christian E. Elger, and Juergen Fell, "Cross-Frequency Coupling Supports Multi-Item Working Memory in the Human Hippocampus," *Proceedings of the National Academy of Sciences of the United States of America* 107 (2009): 3228–33.

Ryan T. Canolty and Robert T. Knight, "The Functional Role of Cross-Coupling," *Trends in Cognitive Sciences* 14 (2010): 506–15.

Paul Sauseng, Birgit Griesmayr, Roman Freunberger, and Wolfgang Klimesch, "Control Mechanisms in Working Memory: A Possible Function of EEG Theta Oscillations," *Neuroscience & Biobehavioral Reviews* 34 (2010): 1015–22.

CHAPTER 9: INTRODUCTION TO THE HARD PROBLEM

1. Edward N. Lorenz, "Predictability: Does the Flap of a Butterfly's Wings in Brazil Set Off a Tornado in Texas?" (paper presented at the meeting of the American Association for the Advancement of Science, Washington, DC, 1972).

2. A number of prominent cosmologists have written about the apparent fine-tuning of our universe, making it life-friendly. See Paul Davies, *The Goldilocks Enigma* (New York: First Mariner Books, 2006); Luke A. Barnes, "The Fine Tuning of Nature's Laws," *New Atlantis* 47 (Fall 2015): 87–97, . http://www.thenewatlantis.com/publications/the-fine-tuning-of-natures-laws; Geriant F. Lewis and Luke A. Barnes, *A Fortunate Universe: Life in a Finely Tuned Cosmos* (New York: Columbia University Press, 2016).

3. Steven Laureys and Giulio Tononi, eds., *The Neurology of Consciousness* (Amsterdam: Academic Press, 2009). Tononi (along with Gerald Edelman) has promoted the plausible idea that high complexity and consciousness require large regions of inactive neurons, which act as essential parts of the dynamic patterns of information (see note 2 of chapter 1). The Laureys and Tononi book is critically reviewed by Edward Kelly, who emphasizes that its excellent coverage of mainstream neuroscience serves to remind us of how far we are from even asking the right questions about the hard problem; in Kelly's words:

Like classical physics of the late 19th century it is a picture of great power and beauty; nevertheless, just as discordant phenomena such as black-body radiation and the photoelectric effect presaged the rise of quantum mechanics, this book, precisely because of its clarity, reveals signs of trouble ahead.

Edward Kelly, "*The Neurology of Consciousness*," *Journal of Scientific Exploration* 26 (2012): 245–57.

4. Proposed levels of consciousness similar to table 9.2 appear in a book by Michio Kaku, a theoretical physicist who based his list largely on multiple interactions with neuroscientists: Michio Kaku, *The Future of the Mind* (New York: Anchor Books, 2014).

5. Raymond Tallis, *Why the Mind Is Not a Computer: A Pocket Lexicon of Neuromythology* (Exeter: Imprint Academic, 2004); Raymond Tallis, *Aping Mankind: Neuromania, Darwinitis, and the Misrepresentation of Humanity* (Durham, UK: Acumen, 2011). Tallis advances strong arguments for the importance of the temporal depth of consciousness when addressing the question of free will.

6. James Gleick, *The Information* (New York: Vintage Books, 2012).

7. Binary numbers are represented by 1s and 0s placed in strings. Larger numbers require more bits, for example, seven bit numbers can represent any base-ten number between 0 (binary 0000000) and 127 (binary 1111111), which is $2^6 + 2^5 + 2^4 + 2^3 + 2^2 + 2^1 + 2^0 = 64 + 32 + 16 + 8 + 2 + 2 + 1 = 127$. In another example, the base-ten number $16 = 2^4$ in binary is (0010000) or (10000), which is the same thing expressed in a shorter bit string (adding zeros to the left side of any number, base-ten or base-two or any other base, does not change its value, for example, 00993 = 993).

8. Seth Lloyd, *Programming the Universe* (New York: Vintage Books, 2006).

9. Detailed discussion of quantum weirdness is beyond the scope of this book. Fortunately, many excellent nonmathematical treatments have been published over the past several decades; some are listed under notes 3 and 9 of chapter 10 below. My recent favorite is Jim Al-Khalili, *Quantum. A Guide for the Perplexed* (London: Weidenfeld & Nicolson, 2003).

CHAPTER 10: MULTISCALE SPECULATIONS ON THE HARD PROBLEM

1. Daniel C. Dennett, *Consciousness Explained* (New York: Little, Brown, 1991).

2. Will Knight, "What Marvin Minsky Still Means for AI," *MIT Technology Review*, January 26, 2016, https://www.technologyreview.com/s/546116/what-marvin-minsky-still-means-for-ai/ (accessed August 1, 2016).

3. Lee Smolin, *Three Roads to Quantum Gravity* (New York: Basic Books, 2001); Brian Green, *The Hidden Reality* (New York: Vintage Books, 2011).

4. Much of the a priori skepticism about psi may be driven by its apparent conflict

with the prevailing materialist position, but that's precisely what makes it interesting. One can argue that even a one percent chance that some aspect of psi actually exists justifies study at some level, especially if privately funded.

5. Much of SETI's support over the past thirty-six years or so has come from the Planetary Society. While here I have downgraded the chances of SETI success in the foreseeable future, I have also been a member of the Planetary Society since its inception in 1980 and remain mildly supportive of their SETI program.

6. Alwyn Scott, *Stairway to the Mind* (New York: Springer-Verlag, 1995).

7. I have repeatedly contended that descriptions of the brain at large scales should not be regarded as poorly resolved approximations of an underlying microscopic order; rather, different scales offer parallel and complementary views of brain organization. This idea has long been appreciated by complex physical systems scientists, and it is nicely developed specifically for brain networks in the recent book by Olaf Sporns (see endnote 1 of chapter 8). Even in relatively simple systems, multiscale theories are typically required for experimental verification, as in the prominent example of separate micro- and macroscopic versions of Maxwell's equations governing electromagnetic phenomena. Much of early twentieth-century physics was concerned with establishing cross-scale relationships between electromagnetic fields observed at different scales, but the analogous developments in brain science are now only in their very early stages. One of the most original and sophisticated scale-crossing theories of brain function is Lester Ingber's statistical mechanics of neocortical interactions, first published in a prestigious physics journal in 1982. See endnote 6 of chapter 3 and http://www.ingber.com/ for a complete reference list to Ingber's papers; most are available online.

8. Murray Gell-Mann and Seth Lloyd, "Information Measures, Effective Complexity, and Total Information," *Complexity* 2 (1996): 44–52; Seth Lloyd, *Programming the Universe* (New York: Vintage Books, 2007).

9. I have discounted highly speculative concepts like wormholes, which might connect distant regions of space-time.

10. Detailed discussions of quantum mechanics are beyond the scope of this book; I did provide a brief introduction to some of the main ontological issues in Paul L. Nunez, *Brain, Mind, and the Structure of Reality* (New York: Oxford University Press, 2010).

The fanciful Alice-Bill-Bob cartoon of section 10.6 refers to the famous two-slit physics experiment in which turning on a sensor observing one slit somehow influences the nature of the waves or particles passing through both slits. Another especially weird aspect of quantum mechanics is revealed by EPR-related experiments based on *Bell's theorem*. *EPR* stands for the three scientists—Einstein, Podolsky, and Rosen—who introduced a famous thought experiment in 1935 to argue that quantum mechanics is an incomplete physical theory, implying that elementary particles must contain mysterious hidden features that are unknown to quantum mechanics. The basic story goes like this—it is well-established that elementary particles can become *entangled* such that even when they become separated over large distances, experimental observations of one particle is tightly correlated with observed properties of the other particle, apparently even when one particle is located in distant parts of the universe. Bell's theorem shows how the most "obvious" explanation, the proposition that the particles acquire hidden features when entangled, can be tested experimentally.

A classical metaphor that reveals the general ERP idea involves two scientists Lisa and Michael. Suppose two "quantum coins" become *entangled* in some manner and are later flipped in widely separated locations where Lisa and Michael reside. Lisa flips the first coin—it lands on a wood surface and comes up heads. Quantum mechanics then says that the second (distant) coin must come up tails if Michael tosses it on the same wood surface—this result is required by the principle of entanglement. Perhaps you think that this is not so surprising—maybe the coins acquired some hidden features to force this outcome when they first became entangled.

But, this "obvious" answer is wrong. Experiments now show that coin-toss outcomes are not predetermined by some features hidden within each coin. Rather, a definite state of the coins can only be determined after one coin is tossed. When Lisa tosses her coin on wood and obtains a head, Michael's toss of the entangled (twin) coin on wood must yield a tail. Somehow the coins are linked by a strange *nonlocal influence*. Apparently, if Michael tosses first, his coin outcome fixes the outcome that Lisa must obtain with a later coin toss, and vice versa. Einstein cited this "spooky action at a distance" as proof that quantum mechanics must be incomplete. The famous debate between Einstein and Bohr on this subject went on for more than twenty years. Some say that Einstein was proved wrong and Bohr won the debate, but the outcome is more nuanced than that. No classical explanation can solve the mystery; quantum mechanics is inherently nonlocal—we better just get used to it. Many popular books on quantum mechanics have discussed this effect as well as other weird quantum behaviors; here is a partial list:

Nick Herbert, *Quantum Reality* (New York: Anchor Books, 1985).
David Deutsch, *The Fabric of Reality* (New York: Penguin Books, 1997).
Jim Al-Khalili, *Quantum. A Guide for the Perplexed* (London: Weidenfeld & Nicolson, 2003).
Brian Green, *The Fabric of the Cosmos* (New York: Vintage Books, 2004).
Bruce Rosenblum and Fred Kuttner, *Quantum Enigma* (New York: Oxford University Press, 2006).

11. Ibid.
11. Ibid.
12. Ibid.
13. In Bohm mechanics, Schrödinger's equation for a single nonrelativistic particle

is transformed into the following equation in the standard form of Newton's law with the usual classical force $\mathbf{F}_c$ and velocity $\mathbf{v}$, but with an added quantum "correction" involving the function R, the *magnitude* of the particle's quantum wave function,

$$m\frac{d\mathbf{v}}{dt} = \mathbf{F}_c + \frac{\hbar^2}{2m}\nabla\left(\frac{\nabla^2 R}{R}\right).$$

When the particle mass m is very large or the Planck's constant term involving $\hbar$ is small compared to other terms, this equation reduces to Newton's law indicating that mass times acceleration equals force. The Bohm interpretation of quantum mechanics is apparently not favored by most physicists; however, it has at least one major supporting feature—it exposes quantum mechanics as an unequivocal *nonlocal theory* because the "quantum correction" term fails to fall off with distance. I emphasize that this nonlocal feature was *not* introduced by Bohm; rather it was already "embedded" in Schrödinger's equation as discussed in David Bohm and Basil J. Hiley, *The Undivided Universe* (New York: Routledge, 1993).

When *nonlocal influences* were well-established by EPR-related experiments, some "nontraditional" physicists attempted to show that such influences might be employed to send superluminal (faster than light speed) signals between remote locations. While such superluminal signals are evidentially impossible, the advocates did propose some interesting ideas that required some serious thought by physicists to discredit. The story is somewhat analogous to classical attempts to design perpetual-motion machines—many clever ideas were proposed, but all ultimately failed. This is an interesting story, so don't be put off by the flaky title: David Kaiser, *How the Hippies Saved Physics* (New York: Norton, 2011).

14. See the list of books under endnote 10 for this chapter.

15. Geoff Haselhurst, "Quantum Theory: Albert Einstein," On Truth and Reality, http://www.spaceandmotion.com/quantum-theory-albert-einstein-quotes.htm (accessed August 1, 2016).

16. See the list of books under endnote 10 of this chapter.

17. The following edited book contains chapters by authors from varied backgrounds; many take seriously the idea that information may underpin physical reality: Paul Davies and Niels Henrik Gregersen, eds., *Information and the Nature of Reality* (Cambridge, UK: Cambridge University Press, 2010).

INDEX

abiogenesis, 279, 331

action potential, 74, 80–83, 86–88, 96, 107, 126, 141–142, 160–166, 207–211, 214, 236, 249, 272, 288, 311, 328, 331, 338–339, 346, 354

ADHD (attention deficit hyperactivity disorder), 83

AI (artificial intelligence), 9, 23, 30–31, 54, 254, 277–278, 282, 299, 331, 348

agnostic, 18, 45, 347

Al-Khalili, Jim, 361, 363

alpha rhythm. *See under* EEG

ALS (amyotrophic lateral sclerosis), 128, 131

Alzheimer's disease, 6, 9, 15, 32, 98, 118, 128, 156–158, 165, 286–287, 331

amyotrophic lateral sclerosis. *See* ALS

analogues, metaphors, and cartoons, 9, 50, 57, 70, 77, 115, 136, 147, 167–169, 200, 262, 331, 338

android. *See* AI

anesthesia. *See under* consciousness

anthropic principle. *See under* philosophical issues

aphasia, 130–133, 331

Aristotelian idealization. *See under* philosophical issues

artifact. *See under* EEG

artificial intelligence. *See* AI

artificial neural network. *See* AI

astrocyte, 76, 331

atheist, 17–18, 334, 347

attention deficit hyperactivity disorder (ADHD), 83

autism, 9, 15, 77, 83, 106, 163, 278, 331

autonomic system, 74, 81, 106

axon, 75, 80–81, 86–88, 91, 95–96, 103–107, 149, 160–165, 208–209, 227–250, 272, 277, 291, 331, 334–336, 339–355, 359

Baars, Bernard, 12, 351

backward masking. *See* masking

Barabasi, Albert-Laszlo, 357

Barlow, John, 353

Barnes, Luke, 360

Basar, Erol, 353

Bassett, Danielle, 356

Bauby, Jean, 129, 350

Bell's theorem. *See* quantum mechanics, EPR

Berger, Hans, 194, 199, 360

binding by resonance, 8, 266–271, 332

binding by synchrony, 189, 269, 332

binding problem, 80, 193, 203, 233, 247, 261, 266, 271, 332

binocular rivalry, 7, 230–234, 256

biochemistry, 250, 306–308, 313, 317, 327

bipolar disorder, 163

black hole, 309, 312, 314, 332

Blackmore, Susan, 345

blindsight, 145, 332

Bohm, David, 314, 318, 332, 343, 363–364

Bohm mechanics. *See under* quantum mechanics

bottom-up interactions. *See* circular causality

boundary conditions. *See* environmental conditions

brain

brainstem, 57, 74, 77, 80, 89, 119, 122–123, 128–131, 136, 160, 252, 286

Broca's area, 131, 137, 332

cerebellum, 74, 284, 291, 297, 303

CCC (cortico-cortical column), 91–92, 147, 248, 258, 333–334

connectome, 95, 162, 240, 249, 271–272, 291–292, 334, 356

cortico-cortical axon, 86, 91, 107, 162, 249, 272, 291, 346

CSF (cerebrospinal fluid), 209, 223, 333, 335